ANALYTICAL METHODS IN
COMBINATORIAL CHEMISTRY
SECOND EDITION

CRITICAL REVIEWS IN COMBINATORIAL CHEMISTRY

Series Editors

BING YAN
School of Chemistry and Chemical Engineering
Shandong University, China

ANTHONY W. CZARNIK
Department of Chemistry
University of Nevada–Reno, U.S.A.

A series of monographs in molecular diversity and combinatorial chemistry, high-throughput discovery, and associated technologies.

Analytical Methods in Combinatorial Chemistry, Second Edition
Bing Yan and Bin Zhang

Combinatorial and High-Throughput Discovery and Optimization of Catalysts and Materials
Edited by Radislav A. Potyrailo and Wilhelm F. Maier

Combinatorial Synthesis of Natural Product-Based Libraries
Edited by Armen M. Boldi

High-Throughput Lead Optimization in Drug Discovery
Edited by Tushar Kshirsagar

High-Throughput Analysis in the Pharmaceutical Industry
Edited by Perry G. Wang

A Practical Guide to Assay Development and High-Throughput Screening in Drug Discovery
Edited by Taosheng Chen

ANALYTICAL METHODS IN COMBINATORIAL CHEMISTRY

SECOND EDITION

Bing Yan

Bin Zhang

CRC Press
Taylor & Francis Group
Boca Raton   London   New York

CRC Press is an imprint of the
Taylor & Francis Group, an **informa** business

CRC Press
Taylor & Francis Group
6000 Broken Sound Parkway NW, Suite 300
Boca Raton, FL 33487-2742

First issued in paperback 2017

© 2011 by Taylor and Francis Group, LLC
CRC Press is an imprint of Taylor & Francis Group, an Informa business

No claim to original U.S. Government works

ISBN 13: 978-1-138-11668-9 (pbk)
ISBN 13: 978-0-8493-1860-3 (hbk)

Library of Congress Cataloging-in-Publication Data

Yan, Bing.
   Analytical methods in combinatorial chemistry / Bing Yan and Bin Zhang. -- 2nd ed.
      p. cm. -- (Critical reviews in combinatorial chemistry)
   Includes bibliographical references and index.
   ISBN 978-0-8493-1860-3 (alk. paper)
   1. Combinatorial chemistry. I. Zhang, Bin, 1968- II. Title.

RS419.Y36 2011
615'.19--dc22   2010004973

Visit the Taylor & Francis Web site at
http://www.taylorandfrancis.com

and the CRC Press Web site at
http://www.crcpress.com

Contents

Preface

During the preparation of this edition, we were as excited as when we were working on the first edition of this book. After 11 years, chemical library and combinatorial chemistry methods have been developed into mature technologies and merged into various cutting-edge approaches, such as chemical biology, chemical genetics, drug discovery, and nanotechnology. The concepts and technologies have also been integrated into the discovery processes of many industries and scientific disciplines.

Over the past decade, combinatorial chemistry has been gaining strength. It has become an active research field worldwide after its beginnings in the United States. Combinatorial chemistry publications are now mostly from academia, whereas a decade ago they were mostly from industry. All pharmaceutical companies have accepted compound libraries as a main source of screening compounds, although they are now almost entirely outsourced to countries such as China, Russia, and India. Internally they rely heavily on small library synthesis to optimize hit compounds. And applications of combinatorial chemistry approaches in other disciplines, such as those concerned with catalysts, material, and agricultural products, are very successful.

Over the years, there have been significant shifts in emphasis in combinatorial synthesis. This edition reflects such shifts. First, the demand for quality control of every library member stimulated the development of many high-throughput analytical techniques and the associated informatics. We have thus significantly modified and enriched Chapter 6, on high-throughput analytical methods. Second, the chromatographic purification of every compound in a library has become a must. We therefore added a new Chapter 7, focused entirely on high-throughput purification methods. Third, we updated all other chapters with new data, and the new Chapter 8 was added to describe future directions and the associated analytical challenges. We are very optimistic about the next wave of progress in this field in the coming decade.

Bing Yan

Member, St. Jude Children's Research Hospital, Memphis, Tennessee, U.S.A.
Professor, Shandong University, Jinan, China

Bin Zhang

Associate Professor, Shandong University, Jinan, China

Authors

Bing Yan received his Ph.D. from Columbia University with Koji Nakanishi in 1990. He was a postdoctoral fellow at the University of Cambridge from 1990 to 1991, and at University of Texas Medical School in Houston from 1991 to 1993. From 1993 to 1999, he worked at Novartis in the area of Core Drug Discovery Technologies as assistant fellow, associate fellow, and senior associate fellow. From 1999 to 2004, he worked at Discovery Partners International as the director of analytical sciences. From 2004 to 2005, he worked at Bristol-Myers Squibb as a senior principal scientist. Now he is a faculty member at the Department of Chemical Biology and Therapeutics at St. Jude Children's Research Hospital in Memphis, Tennessee, and an adjunct professor at Shandong University. He has published or edited eight books and more than 105 peer-reviewed papers. He is the editor of the book series *Critical Reviews in Combinatorial Chemistry* published by the Taylor & Francis Group. He also serves as the associate editor for the *Journal of Combinatorial Chemistry* published by the American Chemical Society.

Bin Zhang received his Ph.D. from the Institute of Inorganic and Analytical Chemistry and Radiochemistry at Saarland University (Germany) with Professor Doctor Horst P. Beck in 2003. He is currently an associate professor of analytical chemistry in the School of Chemistry and Chemical Engineering at Shandong University, China. His scientific interests include the development of analytical methods in combinatorial chemistry and the characterization and analysis of surface chemically functionalized nanoparticles and their interaction with biological molecules.

chapter one

Analytical issues in combinatorial chemistry

1.1 Combinatorial chemistry

Combinatorial chemistry research has undergone enormous development since the 1990s (Figure 1.1), and the practice of combinatorial chemistry has spread worldwide. These developments have been the subject of many reviews and books (Reviews: Jung and Beck-Sickinger, 1992; Pavia et al. 1993; Felder 1994; Janda 1994; Gordon et al. 1994; Czarnik 1995; Lyttle 1995; Thompson and Ellman 1996; DeWitt and Czarnik 1996; Still 1996; Balkenhohl et al. 1996; Lam et al. 1997; Fenniri 1996; Trautwein et al. 1997; Wentworth Jr. and Janda 1998; Dolle 2002 to 2005; Dolle et al. 2006; Schmatloch et al. 2003; Kenseth and Coldiron 2004; Zhou 2008; Moos et al. 2009. Books: Bannwarth and Hinzen 2006; Jung 2000; Bunin 1998; Gordon and Kerwin 1998; Obrecht and Villalgordo 1998; Terrett 1998; Czarnik and DeWitt 1997; Devlin 1997; Moos et al. 1997; Wilson and Czarnik 1997; Abelson 1996; Chaiken and Janda 1996; Cortese 1996; Epton 1996; Jung 1996; Geysen et al. 1995; Houghten 1995; Merrifield 1993).

Combinatorial chemistry can be defined (Czarnik 1998) as a subfield of chemistry that aims to combine a small number of chemical reagents, in all the combinations determined by a given reaction scheme, to yield a large number of well-defined products in a form that is easy to screen for properties of interest. It can be considered an umbrella term that covers a wide array of topics, including the combinatorial and high-throughput parallel synthesis and all the associated technologies for handling, analyzing, and screening these libraries and obtaining useful molecules from them. Combinatorial chemistry has come to be regarded as a tool rather than an end in itself. Its success is measured in terms of the lead compounds and new chemical entities discovered in various areas, such as molecular recognition, catalysis, drug discovery, materials science, and agrochemical applications.

The integration of combinatorial chemistry and high-throughput (HT) screening has provided a versatile and powerful tool in the drug discovery process. In drug discovery, several factors have catalyzed the dramatic growth of combinatorial chemistry. First, the rapid

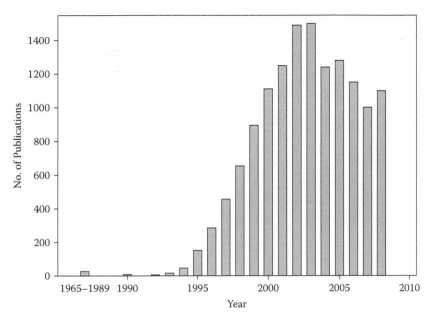

Figure 1.1 Number of publications in combinatorial chemistry based on a search result in SciFinder for the term *combinatorial chemistry*.

developments in genetics and molecular biology can now provide an ever-increasing number of therapeutic target proteins for testing. Second, to meet the need for screening biological targets, significant progress in high-throughput screening technologies has been made since the beginning of the 1990s. It is possible to test more than 100,000 compounds in a period of months. This has made it clear that the synthesis of molecules has become the rate-limiting step in the drug-discovery process. Third, the huge cost and the lengthy process of developing a marketable drug have become formidable, so that all pharmaceutical companies are eagerly searching for more efficient and cost-effective ways of drug discovery. Figure 1.1 shows a literature search using SciFinder for the term *combinatorial chemistry*. Although the use of this term has had a 3% drop in recent years, publications on *parallel synthesis, high-throughput synthesis,* and *compound library synthesis* have increased by the same percentage. The combined research and publications have kept steady for the past eight years. Although no longer in the spotlight or heralded as the savior of the drug industry, combinatorial chemistry is alive and well: actually, combinatorial chemistry science is more prevalent and widespread than ever before (Kennedy et al. 2008).

Combinatorial synthesis is a technology that allows the rapid synthesis of large numbers of compounds with diverse features (structural and physicochemical properties). The term *combinatorial chemistry* used to be more accurately reserved for the synthesis of compound libraries containing an extremely large number of compounds with all possible combination of substituents, to distinguish it from *parallel synthesis*, where the compound libraries consist of a smaller number of individual or discrete compounds; but it is now used indiscriminately to embrace both methodologies. Combinatorial libraries can be made on solid supports, on soluble polymer supports, or in solution.

From the vast new sources of organic compounds, efficient screening machines identify active compounds for defined targets—a process known as *lead generation*. Once active compounds are found, more focused libraries based on those chemical structures provide valuable structure-activity relationship data by way of what is known as *lead optimization*. Note that current lead generation and lead optimization are based only on *in vitro* activity. Physicochemical and other properties are optimized during a later developmental stage. Ideally, compounds with good physicochemical properties should also be selected as lead compounds, and their *in vitro* activity can be optimized in lead optimization.

In the lead identification phase, the goal is to make the most diverse organic compound libraries in the shortest time. The molecular diversity of a compound library means the greatest variation in molecular structure, size, shape, polarity, and polarizability—therefore, with a wide variation in *in vitro* activity and physicochemical properties such as LogP, solubility, permeability, metabolic stability, and bioavailability. The lead selection should be based on all of these criteria. Unfortunately, for technical and philosophical reasons, lead compounds are currently selected on the basis of *in vitro* activity alone.

1.2   Synthesis methods

Few advances in chemistry have had as profound an effect on biotechnology as solid-phase-supported synthesis. First developed by Merrifield (1963) for peptide synthesis, the method has already been adapted to small organic molecules (Fridkin et al. 1966; Leznoff 1974; Crowley and Rapoport 1976; Leznoff 1978; Frechet 1981; Hodge 1988; Brown 1997) and is now routinely used in laboratories worldwide. Although SPOS was continuously studied by several academicians in the 1970s and 1980s, the technology flourished only after the seminal publications of Bunin and Ellman (1992) and of De Witt et al. (1993).

Solid-phase chemistry has advantages that make it an extremely important combinatorial methodology. Since the synthesis is conducted on a solid support, the removal of excess reagents and byproducts is accomplished easily by a copious washing of the polymer. A large excess of reagents can be added and then washed away after the reaction terminates. This allows each reaction step to be driven very close to completion. If, on the other hand, the reaction efficiency were to be much less than 100%, then a fraction of unreacted chains would remain incomplete at the end of each reaction step. Multiply this by any reasonable number of steps, and the resulting combinatorial library will contain a large portion of undesired products.

Two formats for library synthesis are high-throughput parallel synthesis on polymer supports or in solution, and combinatorial synthesis on polymer support using methods such as "split-and-pool" synthesis (also known as *split synthesis*). Parallel synthesis makes discrete compounds and, if they are pure, provides substances for unambiguous testing. The number of compounds and the diversity made through this method are, however, limited. Split synthesis, on the other hand, is a promising technique for covering much more chemical diversity. It is capable of making a library of millions of compounds, with each bead containing only one compound. In a sense, split synthesis also makes discrete compounds at a single-bead level. Because of the huge number of compounds, it is not practical at present to assay discrete compounds released from every bead. Future development in the miniaturization of HT screening will enforce this format. At present, compounds from split libraries are usually assayed in the form of mixtures. The screening of pooled compounds in solution, however, may give ambiguous results. On-bead screening has also been combined with tagging schemes in identifying lead compounds.

After years of widespread practice of SPOS, difficulties associated with this technology became more evident. First, the synthetic methods required to make organic molecules on a polymer support have not been optimized. The shortage of SPOS literature, the insufficient knowledge of solid support and its effect on SPOS, and the lack of effective ways of monitoring many diverse organic reactions on support have significantly prolonged the optimization of combinatorial reactions. Further problems include the following:

- Reactions are frequently thought to be retarded on solid phase owing to the lack of understanding of solid-phase reaction kinetics and fundamental effects of support on organic reactions.
- Heterogeneous reagents are typically less effective.
- Resin supports display significant solvent-dependent swelling and shrinking properties, which can severely affect reaction rates and site accessibility.
- The optimal solvent for resin swelling may not be the optimal solvent for the desired reaction.
- There is an attachment step and a cleavage step.

For these reasons, the size of the typical library has been decreasing since the end of the last century, while there has been an upswing of solution-phase synthesis.

Even though a lot of work is needed to perfect the SPOS methodology, SPOS is still the foundation of the now most widely practiced form of combinatorial chemistry and remains the method of choice for most combinatorial library synthesis. In addition to solid-phase synthesis, the synthesis of small-molecule combinatorial libraries has also been pursued using both liquid-phase (on soluble polymers see Bayer and Mutter 1972; Bonora et al. 1990; Han et al. 1995) and solution-phase chemistry (Storer 1996; Garr et al. 1996; special issue of *Tetrahedron* 1998).

In liquid-phase combinatorial synthesis (LPCS), the key is a soluble, linear homopolymer (polyethylene glycol monomethyl ether (MeO-PEG)), which also serves as a terminal protecting group for the library of compounds synthesized. Two properties inherent in this homopolymer's structure provide the necessary elements for it to be attractive in a combinatorial format. First, with its helical structure, MeO-PEG has a strong propensity to crystallize; thus, as long as the polymer remains unaltered during the construction of the library, purification by crystallization can be accomplished at each stage of the synthesis. Second, MeO-PEG has remarkable solubilizing effects in a variety of aqueous and organic solvents. This property can be used to advantage if the homopolymer is treated as a reagent and used in large excess. Another virtue of the polymer's good solubility is that all manipulations in the LPCS, including split synthesis, can be carried out under homogeneous conditions. Furthermore, reaction progress can be monitored by NMR. The drawback is that intermediate separation and purification are not as simple as in solid-phase synthesis.

Solution-phase combinatorial synthesis has undergone some recent progress. This field is basically an amalgam of synthetic/medicinal chemistry, parallel processing, and laboratory automation. A wide range of reactions is potentially available for solution-phase library generation. In contrast to solid-phase synthesis, no additional steps for attachment to or detachment from supports are required, and thus only the particular reaction of interest needs be developed. The reactions and techniques to be employed in solution-phase library generation are already familiar to all organic chemists. Because the chromatographic purification is impossible in high-throughput synthesis, intermediate purification and removal of excess reagent and solvent are the major hurdles. These make the multistep synthesis nearly impossible. In "split-and-pool" libraries, each bead contains a compound that is physically separated, making tagging to code for the active bead possible, something not possible for solution-pooled compounds.

While combinatorial chemistry matured, new synthetic strategies and techniques, such as multicomponent reaction (MCR) (Hulme and Gore 2003; Habashita et al. 2006; Nishizawa et al. 2007; Liddle et al. 2008), microwave synthesis (Gedye et al. 1986), fluorous synthesis (Studer et al. 1997), click chemistry (Kolb et al. 2001), diversity-oriented synthesis (DOS)

(Schreiber 2000), and flow chemistry (Jas and Kirschning 2003) were born, fine-tuned, and eventually incorporated into combinatorial chemistry to expand the synthesis of novel chemical entities and further explore the realm of molecular diversity. This book addresses both solid-phase and solution-phase methods and compound libraries generated from any method.

1.3   Analytical challenges

The rapid evolution of combinatorial chemistry and the steady state activities in this field constantly pose challenges for analytical technologies and methodologies. Challenges facing analytical chemists involve at least four major areas.

1.3.1   Properties of solid supports

Although there are many advantages to carrying out organic synthesis on solid support, the effects range from basic swelling and solvent compatibility to site-site interaction in the matrix and alteration of reaction kinetics and the entrapped impurities.

Many of these effects have been inadequately studied. The prolonged reaction optimization time, the recent trends towards reduced library size, and a moving away from SPOS can partially be attributed to the lack of knowledge on the effects of solid supports.

1.3.2   Reaction optimization stage

Synthetic route development in combinatorial synthesis remains a great hurdle and a time-limiting factor in most library synthesis projects. It is estimated that reaction development generally can take months, whereas the library synthesis likely takes only weeks. It is not feasible to purify intermediates in multistep synthesis and library products if the libraries are large, so investing work at the route development stage becomes essential. Although it is impractical to characterize all of the products formed during a combinatorial synthesis, significant credibility could be gained by using the same chemical manipulations to synthesize a single representative compound and small trial libraries to measure the success rate. A good quality control (QC) of building blocks and reagents and an analysis of yield, byproducts, and minimization of side reactions are also crucial.

On-support analytical methods are ideal for monitoring reactions at this stage because the chemical changes and the reaction yield on solid support supply the most relevant information. To "cleave and analyze" is time-consuming, laborious, and destructive. Some synthetic intermediates are not even stable enough under cleavage conditions.

1.3.3   Library synthesis stage

After reactions are optimized and trial libraries are rehearsed, a final large library will be generated on polymer supports or in solution. The quality control of this final collection of compounds is a major challenge due to the large number of compounds that are each present in very small amounts. Should QC be done on support or after cleavage? Because most *in vitro* assays are performed in solution, it is advantageous and more relevant to characterize the library in solution. The high-throughput parallel or fast analytical methods are needed for such characterization. Furthermore, the high-throughput methods unavoidably require automatic data interpretation capability, especially for high-throughput NMR methods. It is highly desirable, at least for discrete libraries, to obtain the concentration and purity as the structure is confirmed. In order to make compounds of the same quality as compounds made by conventional methods, all compounds need to be purified and stored for future screenings. High-throughput purification of hundreds of thousands of compounds within a short time posed another serious challenge to analytical chemists.

1.3.4   Lead selection and optimization stage

Should the selection assay be done on support or in solution? Routine HT screening methodology is still the main method for screening discrete compound libraries in solution. If the target biomolecule is soluble, then on-bead screening can be more efficient, and this has been in use. Unfortunately, the utility of on-bead screening is limited by the fact that not all targets are soluble. Additionally, attachment to a solid support can distort the structure of the protein. In-solution screening is clearly more favorable and unambiguous. Another issue is how to balance the unambiguity in single-compound evaluation with the time-efficient, but more complicated, mixture evaluation. The split synthesis actually makes discrete compounds at the level of a single bead. Fewer than one-third of all compounds made on a single low-loading TentaGel bead are enough for a current screening assay (96-well format). As the screening methods become more miniaturized, significantly less material will be needed for evaluation.

An important issue is that the "lead" selection criteria need to be redefined. A new chemical entity must possess a combination of several optimal properties. Taking a drug as an example, a combination of optimal potency, bioavailability, permeability, and physicochemical properties (not to mention stability and feasibility for synthesis and formulation) is required. It would be improper to select leads based only on *in vitro* activity and to disregard compounds with good physicochemical characteristics and bioavailability, but with low scores in the *in vitro* activity assay.

References

Abelson, J. N. *Combinatorial Chemistry*. 1996. Academic Press, San Diego.

Balkenhohl, F., Bussche-Hunnefeld, C., Lansky, A., Zechel, C. 1996. Combinatorial synthesis of small organic molecules. *Angew. Chem. Int. Ed. Engl.* 35, 2288.

Balkenhohl, C., von dem Bussche-Hunnefeld, F., Lansky, A., Zechel, C. 1997. Combinatorial chemistry. *Fresenius J. Anal. Chem.* 359, 3.

Bannwarth, W., Hinzen, B. *Combinatorial Chemistry*. 2006. 2nd edition. Wiley-VCH, Weinheim.

Bayer, E., Mutter, M. 1972. Liquid phase synthesis of peptides. *Nature* 273, 512.

Bonora, G. M., Scremin, C. L., Colonna, F. P., Garbesi, A. 1990. HELP (high efficiency liquid phase) new oligonucleotide synthesis on soluble polymeric support. *Nucleic Acids Res.* 18, 3155.

Brown, R. Recent developments in solid-phase organic synthesis. 1997. *Contemp. Org. Syn.* 4, 216.

Bunin, B. A. 1998. *The Combinatorial Index*. Academic Press, San Diego.

Bunin, B. A., Ellman, J. A. 1992. A general and expedient method for the solid-phase synthesis of 1,4-benzodiazepine derivatives. *J. Am. Chem. Soc.* 114, 10997.

Chaiken, I. M., Janda, K. D. 1996. *Molecular Diversity and Combinatorial Chemistry: Libraries and Drug Discovery*. American Chemical Society, Washington, DC.

Cortese, R. 1996. *Combinatorial Libraries: Synthesis, Screening and Application Potential*. Walter de Gruyter, New York.

Crowley, J. I., Rapoport, H. 1976. Solid-phase organic synthesis: Novelty or fundamental concept? *Acc. Chem Res.* 9, 135.

Czarnik, A. W. 1995. Combinatorial organic synthesis using Parke-Davis's Diversomer method. *Chemtracts – Org. Chem.* 8, 13.

Czarnik, A. W. 1998. Combinatorial chemistry: What's in it for analytical chemists? *Anal. Chem.* 378A.

Czarnik, A. W., DeWitt, S. H. 1997. *A Practical Guide to Combinatorial Chemistry*. American Chemical Society, Washington, DC.

Devlin, J. P. 1997. *High Throughput Screening: The Discovery of Bioactive Substances*. Marcel Dekker, Inc., New York.

DeWitt, S. H., Czarnik, A. W. 1996. Combinatorial organic synthesis using Parke-Davis's Diversomer method. *Acc. Chem. Res.* 29, 114.

DeWitt, S. H., Kiely, J. S., Stankovic, C. J., Schroeder, M. C., Reynolds, D. M., Ravia, M. R. 1993. "Diversomers": An approach to nonpeptide, nonoligomeric chemical diversity. *Proc. Natl. Acad. Sci. U.S.A.* 90, 6909.

Dolle, R. E. 2002. Comprehensive survey of combinatorial library synthesis: 2001. *J. Comb. Chem.* 4, 369.

———. 2003. Comprehensive survey of combinatorial library synthesis: 2002. *J. Comb. Chem.* 5, 693.

———. 2004. Comprehensive survey of combinatorial library synthesis: 2003. *J. Comb. Chem.* 6, 623.

———. 2005. Comprehensive survey of combinatorial library synthesis: 2004. *J. Comb. Chem.* 7, 739.

Dolle, R. E., Le Bourdonnec, B., Morales, G. A., Moriarty, K. J., Salvino, J. M. 2006. Comprehensive survey of combinatorial library synthesis: 2005. *J. Comb. Chem.* 8, 597.

Egner, B. J., Bradley, M. 1997. Analytical techniques for solid-phase organic and combinatorial synthesis. *Drug Disc. Today* 2, 102.

Epton, R. 1996. *Innovation and Perspectives in Solid Phase Synthesis and Combinatorial Libraries.* Mayflower Scientific Limited, Birmingham.

Felder, E. R. 1994. The challenge of preparing and testing combinatorial compound libraries in the fast lane, at the front end of drug development. *Chimia* 48, 531.

Fenniri, H. 1996. Recent advances at the interface of medicinal and combinatorial chemistry: Views on methodologies for the generation and evaluation of diversity and application to molecular recognition and catalysis. *Curr. Med. Chem.* 3, 343.

Fitch, W. L. 1998. Analytical methods for quality control of combinatorial libraries. *Annual Reports in Combinatorial Chemistry and Molecular Diversity* 1, 59.

Fitch, W. L. 1999. Analytical methods for quality control of combinatorial libraries. *Mol. Diversity* 4, 39.

Frechet, J. M. J. 1981. Synthesis and applications of organic polymers as supports and protecting groups. *Tetrahedron* 37, 663.

Fridkin, M., Patchornik, A., Katchalski, E. 1966. Use of polymers as chemical reagents. *J. Am. Chem. Soc.* 88, 3164.

Gallop, M. A., Fitch, W. L. 1997. New methods for analyzing compounds on polymeric supports. *Curr. Opin. Chem. Biol.* 1, 94.

Garr, C. D., Peterson, J. R., Schultz, L., Oliver, A. R., Underiner, T. L., Cramer, R. D., Ferguson, A. M., Lawless, M. S., Patterson, D. E. 1996. Solution phase synthesis of chemical libraries for lead discovery. *J. Biomol. Screening* 1, 179.

Gedye, R., Smith, F., Westaway, K., Ali, H., Baldisera, L., Laberge, L., Rousell, J. 1986. The use of microwave ovens for rapid organic synthesis. *Tetra. Lett.* 27, 279.

Geysen, H. M., Armstrong, R. W., Kay, B. K. 1995. *Chemical Diversity: The Synthetic Techniques of Combinatorial Chemistry.* American Chemical Society, Washington, DC.

Gordon, E. M., Barrett, R. W., Dower, W. J., Fodor, S. P. A., Gallop, M. A. 1994. Applications of combinatorial techniques to drug discovery. 2. Combinatorial organic synthesis, library screening strategies, and future directions. *J. Med. Chem.* 37, 1385.

Gordon, E. M., Kerwin, J. F. J. 1998. *Combinatorial Chemistry and Molecular Diversity in Drug Discovery.* John Wiley & Sons, New York.

Habashita, H., Kokubo, M., Hamano, S.-I., Hamanaka, N., Toda, M., Shibayama, S., Tada, H., Sagawa, K., Fukushima, D., Maeda, K., Mitsuya, H. 2006. Design, synthesis, and biological evaluation of the combinatorial library with a new spirodiketopiperazine scaffold: Discovery of novel potent and selective low-molecular-weight CCR5 antagonists. *J. Med. Chem.* 49, 4140.

Han, H., Wolfe, M. M., Brenner, S., Janda, K. D. 1995. Liquid-phase combinatorial synthesis. *Proc. Natl. Acad. Sci. U.S.A.* 92, 6419.

Hodge, P. 1988. In *Synthesis and Separations Using Functional Polymers.* Sherrington, D. D., Hodge, P., eds. Chapter 2. Wiley, Chichester.

Houghten, R. A. 1995. *Combinatorial Libraries: The Theory and Practice of Combinatorial Chemistry.* ESCOM, Leiden.

Hulme, C., Gore, V. 2003. Multi-component reactions: Emerging chemistry in drug discovery from xylocain to crixivan. *Curr. Med. Chem.* 10, 51.

Janda, K. D. 1994. Tagged versus untagged libraries: Methods for the generation and screening of combinatorial chemical libraries. *Proc. Natl. Acad. Sci. U.S.A.* 91, 10779.

Jas, G. and Kirschning, A. 2003. Continuous flow techniques in organic synthesis. *Chem. Eur. J.* 9, 5708.

Jung, G. 1996. *Combinatorial Peptide and Nonpeptide Libraries: A Handbook.* VCH, Weinheim, New York.

Jung, G. 2000. *Combinatorial Chemistry: Synthesis, Analysis and Screening.* 1st ed., Wiley-VCH, Weinheim, New York.

Jung, G., Beck-Sickinger, A. G. 1992. Multiple peptide synthesis methods and their applications. *Angew. Chem. Int. Ed. Engl.* 31, 367.

Kennedy, J. P., Williams, L., Bridges, T. M., Daniels, R. N., Weaver, D., Lindsley, C. W. 2008. Application of combinatorial chemistry science on modern drug discovery. *J. Comb. Chem.* 10, 345.

Kenseth, J. R., Coldiron, S. 2004. High-throughput characterization and quality control of small-molecule combinatorial libraries. *Curr. Opin. Chem. Biol.* 8, 418.

Kolb, H. C., Finn, M. G., Sharpless, K. B. 2001. Click chemistry: Diverse chemical function from a few good reactions. *Angew. Chem. Int. Ed.* 40, 2004.

Lam, K. S., Lebl, M., Krchnak, V. 1997. The "one-bead-one-compound" combinatorial library method. *Chem. Rev.* 97, 411.

Leznoff, C. C. 1974. The use of insoluble polymer supports in organic chemical synthesis. *Chem. Soc. Rev.* 3, 65.

Leznoff, C. C. 1978. The use of insoluble polymer supports in general organic synthesis. *Acc. Chem. Res.* 11, 327.

Liddle, J., Allen, M. J., Borthwick, A. D., Brooks, D. P., Davies, D. E., Edwards, R. M., Exall et al. 2008. The discovery of GSK221149A: A potent and selective oxytocin antagonist. *Bioorg. Med. Chem. Lett.* 18, 90.

Lyttle, M. H. 1995. Combinatorial chemistry: A conservative perspective. *Drug Dev. Res.* 35, 230.

Merrifield, R. B. 1963. Solid phase peptide synthesis. I. The synthesis of a tetrapeptide. *J. Am. Chem. Soc.* 85, 2149.

Merrifield, R. B. 1993. *Life during a Golden Age of Peptide Chemistry: The Concept and Development of Solid-Phase Peptide Synthesis.* American Chemical Society, Washington, DC.

Moos, W. H., Hurt, C. R., Morales, G. A. 2009. Combinatorial chemistry: Oh what a decade or two can do. *Mol. Diversity* 13, 241.

Moos, W. H., Pavia, M. R., Kay, B. K., Ellington, A. D. 1997. Annual Reports in Combinatorial Chemistry and Molecular Diversity. ESCOM, Leiden.

Nishizawa, R., Nishiyama, T., Hisaichi, K., Matsunaga, N., Minamoto, C., Habashita, H., Takaoka, Y. et al. 2007. H. Spirodiketopiperazine-based CCR5 antagonists: Lead optimization from biologically active metabolite. *Bioorg. Med. Chem. Lett.* 17, 727.

Obrecht, D., Villalgordo, J. M. 1998. *Solid-Supported Combinatorial and Parallel Synthesis of Small-Molecular-Weight Compound Libraries.* Tetrahedron Organic Chemistry Series, V. 17. Pergamon Press.

Pavia, M. R., Sawyer, T. K., Moos, W. H. 1993. The generation of molecular diversity. *Bioorg. Med. Chem. Lett.* 3, 387.

Schmatloch, S., Meier, M. A. R., Schubert, U. S. 2003. Instrumentation for combinatorial and high-throughput polymer research: A short overview. *Macromolecular Rapid Comm.* 24, 33.

Schreiber, S. L. Target-oriented and diversity-oriented organic synthesis in drug discovery. 2000. *Science* 287, 1964.

Shapiro, M., Lin, M., Yan, B. 1997. On-resin analysis in combinatorial chemistry. In ACS Book: *A Practical Guide to Combinatorial Chemistry*, Czarnik, A. W., DeWitt, S. H. eds., 123.

Still, W. C. Discovery of sequence-selective peptide binding by synthetic receptors using encoded combinatorial libraries. 1996. *Acc. Chem. Res.* 29, 155.

Storer, R. Solution-phase synthesis in combinatorial chemistry: Applications in drug discovery. 1996. *Drug Disc. Today* 1, 248.

Studer, A., Hadida, S., Ferritto, R., Kim, S.-Y., Jeger, P., Wipf. P., Curran, D. P. 1997. Fluorous synthesis: A fluorous-phase strategy for improving separation efficiency in organic synthesis. *Science* 275, 823.

Swali, V., Langley, G. J., Bradley, M. Mass spectrometric analysis in combinatorial Chemistry. 1999. *Curr. Opin. Chem. Biol.* 3, 337.

Terrett, N. K. *Combinatorial Chemistry.* 1998. Oxford University Press, New York.

Tetrahedron. 1998. 54. No. 16. Special issue.

Thompson, L. A., Ellman, J. A. 1996. Synthesis and applications of small molecule libraries. *Chem. Rev.* 96, 555.

Trautwein, A., Bauser, M., Fruchtel, J., Richter, H., Jung, G. 1997. *Nachr. Chem. Tech. Lab.* 45, 158.

Wentworth, Jr. P., Janda, K. D. Generating and analyzing combinatorial chemistry libraries. 1998. *Curr. Opin. Biotech.* 9, 109.

Wilson, S. R., Czarnik, A. W. 1997. *Combinatorial Chemistry: Synthesis and Applications.* John Wiley & Sons, Inc., New York.

Yan, B. 1998. Monitoring the progress and the yield of solid phase organic reactions directly on resin supports. *Acc. Chem. Res.* 31, 621.

Yan, B., Gremlich, H. U. 1999. Role of Fourier transform infrared spectroscopy in the rehearsal phase of combinatorial chemistry: A thin layer chromatography equivalent for on-support monitoring of solid-phase organic synthesis. *J. Chromatogr. B.* 725, 91.

Zhou, Z. J. 2008. Structure-directed combinatorial library design. *Curr. Opin. Chem. Biol.* 12, 379.

chapter two

An examination of the analytical sample: resin support

2.1 Introduction

Before the synthesis of a combinatorial library, the sequence of reactions to be used needs to be validated and optimized to give the highest yield and fewest byproducts. Unlike synthesis in solution phase or on soluble polymers, solid-phase synthesis requires a major effort in reaction validation and optimization. One of the major obstacles for adapting reactions to solid support is the lack of knowledge about the solid support and the effects of the support on organic reactions. The consequences of using suboptimal conditions are poor reaction yields, the formation of side products, and even failed syntheses. This chapter examines properties of solid-phase samples and some analytical studies targeted to understand these properties. It will also examine resin swelling and the effects of swollen resin on solid-phase organic reactions. These solid supports are an important platform on which to build combinatorial libraries as well as analytical samples to be analyzed during the reaction optimization process. The success of many analytical methods relies on the physicochemical and swelling properties of solid supports.

During the past three decades, resin properties have been topics of investigations in the context of solid-phase peptide synthesis (SPPS) (Merrifield 1984; Meldal 1997) and solid-phase organic synthesis (SPOS) (Crowley et al. 1973; Overberger and Sannes 1974; Pittaman and Evans 1973; Kraus and Patchornik 1978; Guyot 1988; Hodge 1997). Although the effects of solid supports on SPOS have not been fully evaluated, the effects on SPPS have been studied more thoroughly. For combinatorial synthesis, the practice of SPOS has undergone a rapid expansion both in scale and in depth. So far, however, insufficient attention has been paid to the effects of resin properties in combinatorial synthesis. As a result, unwanted products and impurities are produced unnecessarily along with target products.

SPOS is carried out predominantly on various beaded polymer resins or on grafted polyethylene surfaces. The former have been the major form of support used in solid-phase combinatorial and parallel synthesis and thus are the focus of this chapter. The reaction conditions and the outcome of a solid-phase reaction are usually different from those of its solution-phase counterpart because of the entirely different microenvironment of the solid supports. The physical and chemical characteristics and chemical compatibility of the solid supports often determine the success of SPOS. Although some of the known effects of resin solvation on SPPS reactions may be useful for the optimization of SPOS reactions, differences between SPPS and SPOS must be noted. In this chapter, examples from both categories will be given in order to clarify properties of the support. Instead of considering all SPOS reactions, this chapter will cite selected examples to illustrate some key properties and effects posed by solid supports. Interested readers are referred to several other reviews on related topics (Merrifield 1984; Meldal 1997; Crowley et al. 1973; Overberger and Sannes 1974; Pittman and Evans 1973; Kraus and Patchornik 1978; Guyot 1988; Hodge, 1997; Scott and Steel 2006; Lee et al. 2007).

Resin-based solid-phase synthesis is actually gel-phase synthesis, because it is the highly solvated resin that provides a reaction matrix. In order to illustrate this, the following sequence of topics will be considered:

- Physical properties of resins
- Resin swelling and solvation
- Effects of the solvated resin on solid-phase reactions

2.2   *Physical properties of resins*

Polystyrene-based resin is the most popular solid support for solid-phase combinatorial and parallel synthesis. Its physical properties have a direct impact on the design of the library, the optimization of reaction conditions, and the synthetic scheme. These properties include

- resin type
- size distribution
- loading and the reactive site distribution
- thermal stability
- chemical stability

2.2.1   *Resin type*

2.2.1.1   *Polystyrene resins*

Two main classes of polystyrene (PS) supports, lightly and highly cross-linked PS resins, have been used in SPOS. The most popular PS support in both SPPS and SPOS is a 1% DVB cross-linked gel-type PS resin prepared by suspension polymerization. Other lightly cross-linked (2–5% DVB) gel-type resins and the highly cross-linked (5–60% DVB) macroporous resins have also been investigated. Lightly cross-linked gel-type res-ins have no permanent porosity, but swell to varying degrees in many organic solvents to form micropores: the spaces between cross-links occu-pied by the swelling solvent. These pores can be detected experimentally by contrast-enhanced small-angle neutron scattering (Hall et al. 1997). These spaces contain solutions of polymer segments, not just solvent. Therefore, they are referred to as *microporous*. After removing the solvent, the solvent-induced porous structure and sometimes also the chemically fixed porosity remain preserved. In *macroporous* resins, permanent pores or holes of various sizes are formed. Beads consist of conglomerates of microspheres (100–200 nm) that look like cauliflower florets, and each microsphere shows smaller nuclei (10–30 nm) fused together. Small pores are distributed between the nuclei.

PS resin is highly hydrophobic. Because of the absent spacer, reactions on PS resin tend to be significantly affected by the hydrophobic PS matrix, which may disfavor reactions involving polar and charged species. For example, reactions involving charged, highly polar reagents or aqueous solutions may be retarded. Such unfavorable situations can be corrected with a phase-transfer catalyst (see 2.4.2).

The influence of resin cross-linking on solid-phase chemistry has been investigated recently (Rana et al. 2001). They investigated resins with 0.3 to 6% DVB cross-linking. As expected, the swelling is increased for lower cross-linked resins, and the reaction rate (cleavage of the dye methyl red and a Suzuki reaction) is decreased as the cross-linking increases. Of course, the swelling/reaction solvents also play a role. Interestingly, dur-ing a total synthesis of Kawaguchipeptin B, the product purity is increased as the cross-linking increases.

2.2.1.2 PS-PEG resins

PS-PEG (polyethylene glycol) resins differ from PS resins in that they contain predominantly (up to 70%) PEG spacers (50–60 ethylene oxide units) in their composition (Rapp 1996; Zalipsky et al. 1994). The chemical structures of PS and PS-PEG resins are shown in Figure 2.1. PEG units are immobilized on functionalized PS resin by means of, for example, anionic graft copolymerization. A PS resin bead with a diameter of 44 µm shows an increase in its diameter to 80 µm after being grafted with PEG to the extent that the PEG/PS ratio is 3/1 by weight. PS-PEG shows almost no solvent preference for resin swelling (Figure 2.2). The reason for this is the good miscibility of PEG chains with a majority of solvents. A structural account of the good solvent compatibility of PS-PEG resins was given by Meldal (1997). Because of the vicinal arrangements of carbon-oxygen bonds throughout the chain, PEG assumes helical structures with gauche interaction between the polarized bonds. There are three possible

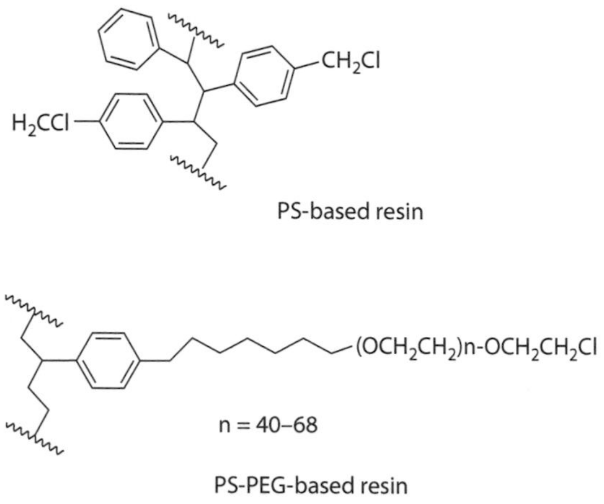

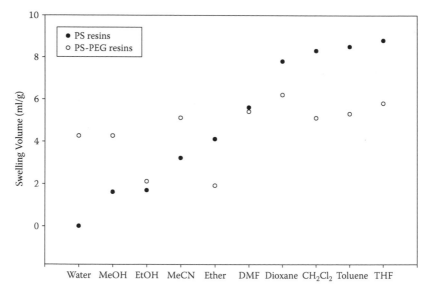

Figure 2.1 Comparison of the chemical structures of PS- and PEG-PS-based resins.

Figure 2.2 Swelling property of PS and PS-PEG resins in different solvents. Data are from Rapp (1996).

helical arrangements of PEG that allow the low-energy gauche conformations of vicinal carbon-oxygen bonds. One is largely hydrophobic with the oxygen atoms in the interior of the helix, the second is of intermediate polarity, and the third is quite hydrophilic, exposing the oxygen atoms to the environment. The PEG molecule thus has an amphipathic nature and is solvated well both by polar and nonpolar solvents.

The properties of the PEG chains therefore determine the mechanical and physicochemical behavior of the resin. Because the reactive sites in TG resins are located at the end of long, flexible spacers and are therefore well separated from the PS backbone, they are less affected by the hydrophobic PS matrix.

2.2.1.3   Other resin and nonresin supports

Kieselguhr (or diatomaceous earth), a porous, inert, inorganic, and highly heterogeneous material prepared as granules from natural sediments, has been used as a matrix for dimethylacrylamide polymerization to afford a resin that was stable to flow (Atherton et al. 1981).

Polypropylene shaped into arrays of pins or small caps attachable to pins (Bray et al. 1990) can be modified by grafting. Next, the surface is polymerized to form a polydimethylacrylamide coating. These rigid pins have been used for the simultaneous synthesis of up to 96 peptides. Microtube® from IRORI (San Diego, CA) is a similar surface-modified polymer support used for SPOS (Li et al. 1998). Several other materials have been derivatized and used for solid-phase synthesis, including beads made from sintered polyethylene, cellulose fibers (cotton, paper, Sepharose, and LH-20), core-shell-type resins (Cho et al. 2002), grafted macroporous polymer monolithic disk (Tripp et al. 2001), Rasta resins (Hodges et al. 2000; Lindsley et al. 2000), JandaJel based on polytetrahydrofuran cross-linkers (Toy and Janda 1999; Reger and Janda 2000; Vaino et al. 2000; Gambs et al. 2003), polymethacrylate, controlled pore glass, and silica. Recent developments in this area were reviewed by Yu and Bradley (2002).

2.2.2   Resin bead size distribution

Resin beads vary in diameter. The most commonly used sizes are 100–200 mesh (150–75 µm) and 200-400 mesh (75–37 µm); large uniform-sized polymer beads are desirable for "one-bead-one-compound" applications in the combinatorial synthesis of compound libraries (Zhu et al. 2006). In general, PS-PEG resins have a more uniform size distribution than PS resins (Lee et al. 2005). However, under proper solvation conditions the variation in 100–400 mesh resin does not show detectable effects on reactivity and kinetics.

Macro PS-PEG resin (400–800 µm) is now commercially available and increases the capacity/bead to 100 or 1000 fold (nmol/bead), whereas the regular PS-PEG resin beads have diameters of 90–130 µm and a 50–200 pmol/bead capacity. Recently, highly condensed and miniaturized assay formats have made the screening of compounds from the individual bead possible. With the introduction of printing techniques, protein-binding assays can now be performed on a slide containing approximately 10,000 compound spots (Hergenrother et al. 2000; MacBeath et al. 1999). Each 250-μm spot contains only 5 pmol of compound. The load from a single macrobead can be made into a stock solution that can be used for hundreds of assays. The compounds synthesized in the macrobeads are also enough for analytical studies such as HPLC and NMR on the materials released from a single macrobead.

Uniform size and loading are important for biological and other functional assays on resin. A study of thiol loading on an individual resin bead has been done (Freeman and Howard 2001), with a broad range of size and mass/bead distributions. The quantitative measurement of the thiol loading also showed a broad distribution.

2.2.3 Loading and the intraresin site distribution

Site distance and concentration have been studied using electron paramagnetic resonance (EPR) spectroscopy (Marchetto et al. 2005). Being a dynamic system, the site distance in resin expands from 17 to 170 Å. Calculated from such a short site-site distance, the site concentration can be as high as 550 mM.

Measured as molar equivalents of reactive groups per gram of resin, the loading capacities are typically 0.5–2.0 mmol/g and 0.2–0.6 mmol/g for PS and PS-PEG resins, respectively. Higher-loading resins have been made in both categories. The loading on grafted crowns or Microtubes depends on the surface area and, therefore, these have a low loading value. The most popular method for determining the loading capacity is to convert a functional group into an Fmoc derivative and then cleave the Fmoc groups for quantitation. Sometimes elemental analysis is used if the resin contains elements such as N, Cl, Br, S, and I (Frechet and Schuerch 1971; N'Guyen and Boileau 1979; Frechet et al. 1979; Mitchell et al. 1978; Forman and Sucholeiki 1995; Yan et al. 1998). Grafted polymers such as crowns and MicroTubes cannot be quantitated by the elemental analysis method, but their loading can be determined using the Fmoc-derivatization or electrochemical methods (Li et al. 1998).

Although manufacturers report the loading capacities of commercial resins and information on the method used for loading determination, the loading value of a resin can change over time because the reactive groups are not inert to storage and handling. Therefore, it is critical to determine the resin loading before the synthesis and to use the same batch of resin throughout the entire process of chemistry development and library synthesis.

Because resins are uniformly functionalized (see below), the relative amounts of reactive groups on the bead surface and its interior can be estimated by volume calculations. If the "depth" of the surface is defined as 0.1 μm, the total amount of reactive sites on the surface is about 0.6% of the total reactive sites for a bead with a diameter of 100 µm. Therefore, more than 99% of these reactive sites are in the interior of the resin bead.

Autoradiography of the microtomed section of resin containing radiolabeled peptide showed random site location (Merrifield 1969). Furthermore, the uniform distribution of reactive sites extends to beads of various sizes. For example, it has been found that resin beads labeled with radioactive peptide showed constant specific activity even though differing in diameter by a factor of 5 (Crowley et al. 1973).

Recently various microscopic methods have been used to map the functional group distribution in single resin beads. Reports using fluorescence microscopic analysis of Rhodamine-labeled beads concluded the nonuniformity distribution of functional groups in polystyrene and TentaGel resins (McAlpine and Schreiber 1999; McAlpine et al. 2001). However, using fluorescence microscopy techniques is tricky. Fluorescence self-quenching or re-absorption was observed in bulk resin suspensions (Scott and Balasubramanian 1997), single-bead fluorescence spectroscopic measurements (Yan et al. 1999), and microscopic study (Egner et al. 1997). IR microscopy (Rademann et al. 2001), scanning confocal Raman spectroscopy (Kress et al. 2001, 2003), and two photon microscopy (Ulijn et al. 2003) confirmed the uniform distribution of functional groups in polystyrene, TentaGel, and PEGA resin beads. IR and Raman microscopy give more conclusive results because they do not have the quenching problem that is associated with fluorescence microscopy.

2.2.4   Thermal stability

Because many organic reactions require high temperatures, the thermal stability of resin beads is an important issue for SPOS. Dry resin exists in a glassy state, and an increase in temperature will transform the resin into a rubbery state. The temperature characterizing this transition is called the glassy temperature (T_g), and T_g values for polystyrene resins copolymerized with 5, 10, and 20% DVB are 102, 116, and 133°C, respectively (Sanetra et al. 1985). Polystyrene-based resins used for SPOS are commonly 1% DVB cross-linked PS, which should have a T_g of < 100°C. In highly cross-linked resins, the glassy transition virtually disappears, and the specific heat continuously increases with increasing temperature up to the threshold of thermal decomposition.

It is unknown whether there is any effect of the glassy/rubbery state transition on the solvation status of resin at high temperatures and on the kinetics and efficiency of the resin-supported organic reactions. Some high-temperature reactions have been carried out well above the T_g, at 130°C, for example (Sun and Yan 1998), suggesting that this transition does not affect polymer-supported organic reactions. However, quantitative studies of the temperature effects are needed to ensure that other high-temperature organic reactions can be transferred to solid supports. On the other hand, organic reactions sometimes need to be carried out at low temperatures, and the effect of resin structure and solvation at very low temperatures needs further study. Finally, it has been noted that another type of rubbery state can be obtained by swelling the resin with solvent and then drying it so that only a small amount of solvent remains (Wieczorek et al. 1982).

2.2.5   Chemical stability

The DVB cross-linked polystyrene matrix is very stable. However, the functional groups on the resin are not. More studies are needed to completely understand their stability. For such studies, more quantitative analytical methods need to be developed. From single-bead FTIR studies, it has been found that trityl chloride resin is not stable, and there always exists a low level of hydrolyzed product. In aldehyde quantitation studies, it has been found that the loading of aldehyde groups declines slowly over time. Other reactive functional groups may behave similarly during storage.

2.3   Resin swelling and solvation

Gel-type (low cross-linked PS) resins swell to two to six times their original volume depending on the solvent used. In this type of resin, solvation is crucial in forming a gel-phase for synthesis. The swelling of macroporous resins is more complicated, because they contain both pores and a polymer phase. The pores fill with solvent, and the polymer phase may swell as in gel-phase resins, except that the cross-link density is higher. Fluorescent molecules attached to resins that were swollen with various solvents showed a solvatochromic shift of the fluorescence emission that correlated qualitatively with the solvent properties (Shea et al. 1989a).

The mutual interactions between the solvent and resin bead result in significant effects on both. On the one hand, as beads swell, their reactive chains become more mobile, so that their reactivity and their influence on the accessibility of incoming reagents change. On the other hand, the mobility of solvent or reagent molecules is greatly reduced when they enter the bead microenvironment. These two aspects are further explained below.

2.3.1   Effects of swelling on resin

Regen (1975) studied the influence of solvent on the motion of molecules "immobilized" on PS matrices. The influence of solvent on the rotational motion of 2,2,6,6-tetramethyl-4-piperidinol-1-oxyl covalently bound to cross-linked PS, silica, and a PS ion-exchange resin has been examined by electron paramagnetic resonance spectroscopy. The degree of swelling of the polymer supports, as defined by the solvent employed, is an important factor in determining the mobility of the attached spin label. In the investigation, it was found that PS (Diagram 2.1) underwent extensive swelling in carbon tetrachloride with an accompanying increase in motion of the nitroxide moiety. When it was swollen in ethanol, slow nitroxide tumbling and little solvent uptake were observed (Figure 2.3). Examination of the influence of solvent on the ion-exchange resin (Diagram 2.1) revealed exactly the reverse effects. Only ethanol and 2-propanol swelled the ion-exchange resin significantly, and only with these solvents was a large increase in the motion of the spin label observed (Figure 2.3). Evidently, the introduction of charged sites along the polymer backbone makes the lattice more compatible with ethanol and less compatible with carbon tetrachloride.

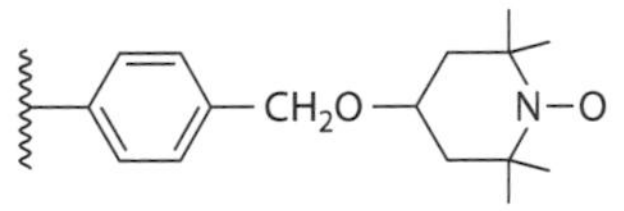

nitroxide-PS-2% DVB

(a)

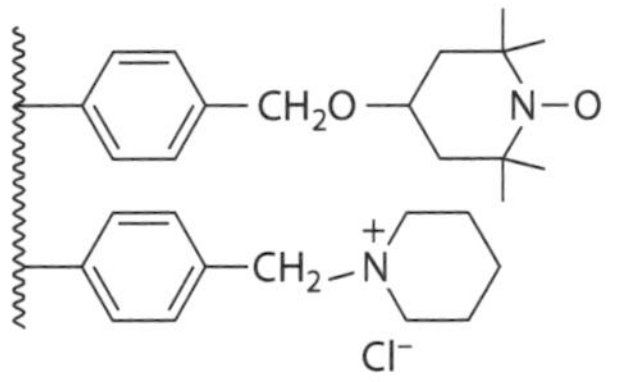

nitroxide-PS-2% DVB ion-exchange resin

(b)

Diagram 2.1

Movement restrictions and the chain conformation in various swollen cross-linked PS gels were studied by ^{1}H NMR line-shape analysis (Schneider et al. 1985). The MAS ^{1}H NMR spectra of PS with 2–10% DVB or ethylene dimethylacrylate, swollen to equilibrium in CCl_4, $CDCl_3$, or benzene, were measured. The spectra were analyzed as a folding of a narrow line-shaped function approximated by the MAS NMR spectrum and as an orientation-dependent dipolar broadening function. As indicated by the NMR spectra, the rate of conformational transition depended on solvent, temperature, and degree of cross-linking, but remained very rapid even at 10% cross-linking. It has been reported that the resin structure, especially the tether length, is the dominant factor influencing ^{1}H NMR linewidths, with the solvent playing only a secondary role (Keifer 1996).

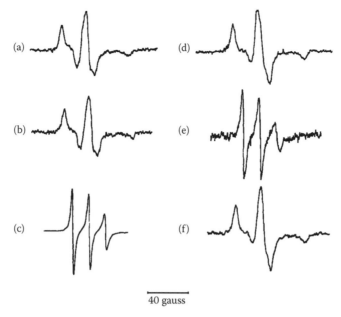

40 gauss

Figure 2.3 EPR spectra of (a) 1 in the dry state, (b) 1 swollen in ethanol, (c) 1 swollen in carbon tetrachloride, (d) 2 in the dry state, (e) 2 swollen in ethanol, (f) 2 swollen in carbon tetrachloride. (Reproduced with permission from *J. Am. Chem. Soc.* 1975, 97, 3108–3112. Copyright© 1975 American Chemical Society.)

Solvation, which is crucial for SPOS, is particularly important for PS resins. Although ^{1}H NMR line width of PS-PEG resins varies little with solvent selection (Figure 2.2), the reactivity of PS-PEG resins may be strongly influenced by the solvent, since the compatibility of the resin and the solvent with reagent and the transition state are also important. For example, the reaction rate of PS-PEG resins is not always higher than PS resins under identical conditions (Li and Yan 1998).

A recent ^{19}F study of seven different resins in four different solvents supported the conclusion that flexible resin structure and better swelling will leave the molecule mobile and give a better NMR spectrum (Mogemark et al. 2004). The best spectral quality in all four solvents was seen in TentaGel and ArgoGel resins (Figure 2.4). They are polystyrene-based resins that are grafted with poly(ethylene glycol) chains (PEG) on the hydrophobic core. They swell well in both polar and nonpolar solvents, and the attached molecules possess a high degree of mobility. NovaGel is also a polystyrene PEG-grafted copolymer. It has a solvent-swelling property similar to those of TentaGel and ArgoGel. However, the location of functional groups is not at the end of PEG chains, being situated instead on the hydrophobic polystyrene core. Except in DMSO, the ^{19}F NMR spectra of this resin is not good. More rigid polystyrene and macroporous

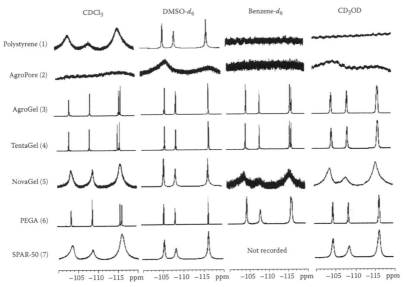

Figure 2.4 ^{19}F NMR spectra of seven resins recorded in CDCl$_3$, DMSO-d_6, benzene-d_6 and CD$_3$OD. (*Org. Biomol. Chem.* 2004, 2, 1770–1776, reproduced by permission of the Royal Society of Chemistry.)

ArgoPore resins gave poor spectra. PEGA and APAR-50 resins are acrylamide-based resins. Since PEGA is polyacrylamide cross-linked with PEG, it gives relatively better spectral quality than do SPAR-50 resins with no PEG in the structure.

2.3.2 Effects of swelling on reagent and solvent molecules

Regarding the effect of support on small molecules, the limited diffusion of reactant or reagent molecules into the bead interior and the reduced mobility of small molecules within the bead are particularly important.

If diffusion of soluble molecules into the active sites of the beads is rate limiting, the result can be the supported reactant displaying significant size selectivity. This is illustrated by the work of Grubbs and Kroll (1971) on the hydrogenation of olefins. A catalyst was prepared by equilibrating 2% cross-linked polystyrene beads in which 8% of the repeat units were benzyldiphenylphosphine residues with a twofold excess of tris-(triphenylphosphine)-chlororhodium(I). When used as a catalyst with benzene as the solvent at 25°C under 1 atm of hydrogen, the relative rates of hydrogen uptake, with hexene arbitrarily set at 100, were cyclohexene 39; cyclooctene 15; cyclododecene (cis-and trans-mixture) 8.8; and Δ^2-cholestene 1.2. For a similar soluble catalyst, the relative rates of hydrogen uptake, again relative to hexene arbitrarily set at 100, were all in the range of 71–100. Therefore, when using a supported catalyst, the rate of reduction is inversely correlated to the molecular size of the olefin.

In the example above, the effect of the solid support on reaction rate is particularly pronounced because most of the catalytic sites likely involve two or more polymer-supported phosphine ligands, which will serve as extra cross-links and restrict diffusion through the matrix. Furthermore, many of the active sites will be on these cross-links. In most other reaction systems, access to the active sites is less of a problem. An example was shown in a single-bead FTIR study of the surface reaction rate vs. the rate in the bead interior for a series of esterification reactions (Yan et al. 1996a). No difference in rates on the bead surface and in its interior was found for those reactions. However, when the reaction is slowed down by incompatibility or other problems such as an aqueous reaction on a polystyrene resin, a different reaction rate on the bead surface and in the bead interior may occur.

Solvent and reagent molecules, when entering the resin bead, undergo changes in their mobility and, perhaps, reactivity. Ford and co-workers (1984) studied the ^{13}C spin-lattice relaxation times (T_1) and NOEs of toluene in cross-linked polystyrene gel beads 16–18% ring substituted with benzyltri-n-butylphosphonium chloride groups. The ^{13}C spin-lattice relaxation time (T_1) is particularly useful because its measurement requires no isotope labeling, and relaxation by the direct dipolar mechanism is intramolecular, allowing interpretation in terms of rotational correlation times. Toluene inside and outside the bead gave separate ^{13}C peaks for each carbon atom. The T_1 values and rate constants for the exchange of toluene between the sites were measured by selective T_1 experiments with concurrent relaxation and exchange. The T_1 values of all protonated carbon atoms of toluene inside the bead were reduced to 0.05–0.33 times their values in low-viscosity toluene solution. The T_1 results are explained by slower rotational diffusion of toluene in the cross-linked polystyrene gels. The self-diffusion coefficient of toluene in the gel is estimated from the exchange rate constants to be 0.02 times that in toluene solution. Self-diffusion coefficients of toluene in swollen cross-linked PS beads with 5.7–40% cross-linking and in linear PS solution were determined by pulsed-field-gradient spin-echo (PGSE) NMR experiments (Pickup et al. 1986). A comparison of solvent diffusion in the cross-linked and the non-cross-linked systems at similar concentrations shows remarkably close agreement. These results suggest that the diffusion rate in the cross-linked polymer beads is not significantly affected by the presence of cross-linking and is determined mainly by the concentration of solvent in the bead. Self-diffusion coefficients of toluene, acetonitrile, chloroform, and dichloromethane in swollen 1–20% cross-linked PS beads were determined by application of a new model to magnetization-transfer ^{13}C NMR data for exchange of solvent in and out of the beads (Ford et al. 1987). In 1% cross-linked beads, the self-diffusion coefficient of dichloromethane is reduced only by a factor of 2 compared with bulk liquid. The poorest swelling solvent acetonitrile shows a reduction in its self-diffusion coefficient by a factor of 15. By comparing interactions of cyclohexane, tetrahydrofuran, and acetonitrile with swollen PS beads using ^{13}C NMR spectroscopy, it was found that the mobility of solvent molecules was affected by the cross-linking density in the case of well-swelling solvents, whereas little dependence on the cross-linking density was seen for poor solvents (Ogino and Sato 1995). These NMR studies show that different solvents have different self-diffusion coefficients in swollen beads depending on their swelling ability, cross-linking density, and the specific interactions of each solvent with resin. Kurth and co-workers (Wilson et al. 1998) studied solvent and reagent accessibility of PS resin cross-linked with ∝,ω-bis-4(vinylbenzyl) ethers of mono-, di-, tetra-, and hexa-9-ethylene glyco. They found that the swelling of 2, 10, and 20% crosslinked PS-PEG beads was higher or comparable to 2% PS beads. Fluorescence quenching of polymer-bound dansyl groups by Et_3OBF_4 indicated that the most rapid intraresin diffusion of this electrophilic reagent occurred in toluene, and that PS-PEG beads exhibited faster quenching than 2% PS beads in poor-swelling solvents but were slower to quench than 2% PS beads in toluene. It is expected that the diffusion of reagent molecules into resin will depend on the molecular properties of the reagent and the properties of the resin.

Macrobeads are a good choice as microreactors in producing more compounds for screening assays and for structural elucidation. In the case of smaller beads (50–100 μm), diffusion is rarely a limiting process in some solid-phase reactions (Yan et al. 1996b). The diffusion of reagents into beads of diameters larger than 500 μm may be a concern. Diffusion of ionic reagents into solvated polymers with dansyl groups attached quenched the fluorescence and revealed the diffusion rate (Shea et al. 1989b). A method for measuring diffusion of reagents in macrobeads has been published (Groth et al. 2001). The authors studied the diffusion rate of various amino-functionalized resins with a diameter of 300 to 1000 μm. These macrobeads were treated with an acylating reagent, and at various time points beads were taken out and treated with acetaldehyde and chloranil (Vojkovsky 1995), which terminate the reaction and stain the remaining amino groups. A PC scanner was used to scan the image of the beads, making possible the diffusion rate calculation. These authors showed that the diffusion was promoted by increasing temperature, good swelling of the resin, a little reagent, and high concentration. They also showed that the agitation and sonication had no effect on the diffusion rate. The effect of reagent diffusion rate on reaction rate was studied for an esterification reaction using three acids of various sizes (Diagram 2.2) and polystyrene resin beads with diameters of 35 to 560 μm (Figure 2.5) (Yan and Tang 2003). When a small acid was used, a reaction retardation in larger beads can be observed. However, when bulkier acids were used and the reaction rate itself became slow, there was no difference in reaction rate for small and larger beads. In case of slow reaction, diffusion is no longer rate-limiting because the reaction rate becomes a bottleneck. An acylation reaction and a nucleophilic substitution reaction were carried out on polystyrene beads of various sizes (40 to 220 µm). The reactions were found to be not diffusion controlled in these beads (Rademann et al. 2001).

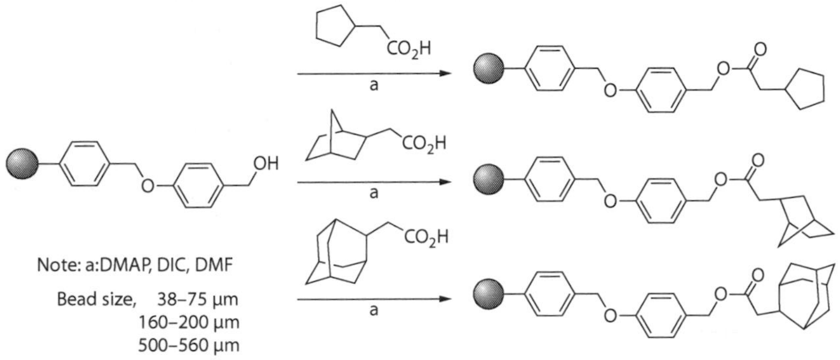

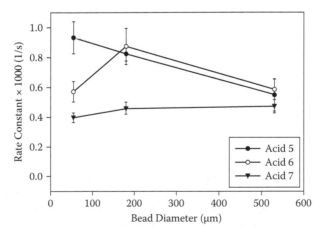

Diagram 2.2

2.3.3 Swelling ability of solvents

Regen (1974) used the spin-labeled technique for examining the mobility of a nitroxide bound to cross-linked PS resin (A in Diagram 2.1 on page 21) in the solvent-swollen state. His results established that the choice of

Figure 2.5 Relationship between the average bead diameter and the reaction rate constant when three acids were used in the reaction (Diagram 2.2). (Reproduced with permission from *Ind. Eng. Chem. Res.* 2003, 42, 5964–5967. Copyright© 2003 American Chemical Society.)

swelling solvent has a substantial influence on the physical nature of the resin-bound nitroxide and that the dominant role of solvent is to define the degree of swelling of the PS lattice (Figure 2.6). A distinct trend is that the rotational correlation time of the spin label decreases as the degree of swelling of the polymer increases. Those solvents that swell PS matrices the most will allow for the greatest mobility of the substrates bound to them. The solvation of resin by single-solvent and mixed-solvent systems can be correlated to the Hildebrand solubility (δ) and hydrogen-bonding solubility (δ_h) parameters (Fields and Fields 1991). A systematic study of the correlation between resin solvation and solvent properties revealed the importance of the sum of solvent electron acceptor (AN) and electron donor (DN) numbers (Cilli et al. 1996). This method measures the capacity of the solvent to act as a Lewis acid or a Lewis base, an acceptor or a donor of electron pairs—AN and DN numbers, respectively. According

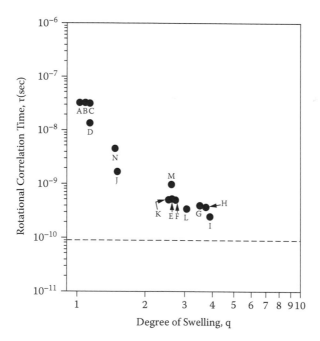

Figure 2.6 Plot of τ as a function of q for 1 swollen in (A) no solvent, (B) ethanol, (C) 2-propanol, (D) acetonitrile, (E) ethyl acetate, (F) N, N-dimethyformamide, (G) carbon tetrachloride, (H) dichloromrthane, (I) benzene, (J) ethanol in benzene (17% v/v), (M) 4% cross-linked swollen in benzene, and (N) 12% cross-linked resin swollen in benzene; the dashed line (------) represents the value of τ measured for a benzene solution of nitroxide on a non-cross-linked polymer. (Reproduced with permission from *J. Am. Chem. Soc.* 1974, 96, 5275–5276. Copyright© 1974 American Chemical Society.)

to this theory, the solvent effect upon a solute molecule is generally considered to be an acid-base type of interaction. AN represents the electrophilic property of the solvent and is related to the relative chemical shift of ^{31}P in triethylphosphine in the particular solvent, with hexane as a reference solvent on the one hand and triethylphosphine oxide–SbCl$_5$ in 1,2-dichloroethane on the other, to which the acceptor of 0-100 has been assigned. In contrast, the DN number is representative of the basic (nucleophilic) property of the solvent molecule and is defined as the molar enthalpy for the reaction of the donor with SbCl$_5$ as reference acceptor in 1,2-dichloroethane. Swelling plotted against (AN+DN) displays variously shaped curves showing the solvation characteristic of each class of resin that, in turn, depend on the nature and amount of the resin-bound peptide sequence (Figures 2.7 and 2.8). In Figure 2.8, each resin has its own characteristically shaped curve. TentaGel and SPAR-50 are broadly solvent compatible and show almost no dependence to (AN+DN) (Malavolta et al. 2002). The occurrence of a maximum solvation region in this plot is important for both SPPS and SPOS methodologies, because it facilitates the selection of the most appropriate solvent system for each step of the synthesis cycle. However, the experimental selection of solvent is now used instead of the theoretical prediction.

Beads must remain swollen throughout the reaction. When the loading of reactive groups is high, the swelling properties can change considerably as the original functional groups are transformed into others, especially when these transformations involve a change in polarity. An

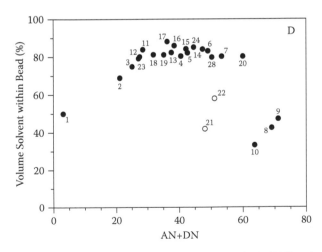

Figure 2.7 Swelling of resin as a function of solvent (AN+DN) value. Symbols 1–25 represent various solvent systems (see Cilli et al. 1996 for details). (Reproduced with permission from *J. Org. Chem.* 1996, 61, 8995. Copyright© 1996 American Chemical Society.)

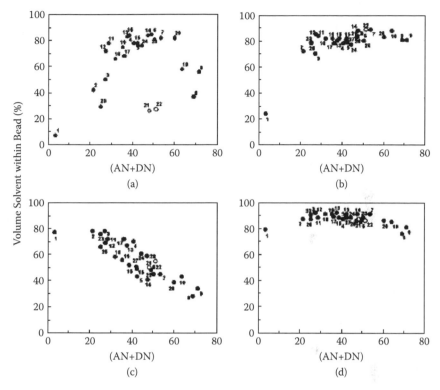

Figure 2.8 Swelling of resins as a function of solvent (AN+DN) values. a: (NANP)$_4$–BHAR, 1.4 mmol/g; b: SPAR-50, 0.6 mmol/g; c: PS–TTEGDA, 0.7 mmol/g; d: TG, 0.3 mmol/g. (Reproduced with permission from *Tetrahedron* 2002, 58, 4383–4394. Copyright© 2002 Elsevier Science Ltd.)

extreme case is the reaction of the Merrifield resin with tertiary amines to give quaternary ammonium salts, where the polymer changes from having relatively nonpolar groups to having ionic groups. Such dynamic changes in resin-swelling properties are discussed next.

2.3.4 Dynamic resin solvation

Swelling is a macroscopic term to describe resin solvation, which is the association of solvent molecules with the resin spacer, the linker, and the bound substrate to create an enlarged bead. The proper solvation of the bound substrate and external reagent is crucial for the reaction to occur, and the success of SPPS and SPOS depends on the accessibility of the growing resin-bound substrate to the reagent. Resin-bound reaction products are expected to have physicochemical properties that differ from the initial resin or compound, especially if the attached compound introduces a component of different polarity. This additional component might require a solvent of different polarity to ensure optimal solvation and, hence, accessibility.

Although studies of the properties of most resin-bound organic substances are still lacking, resin-bound peptides have been well characterized. During the course of a multistep synthesis, both the size of bead and its swelling characteristics will change. Merrifield (1969) showed that the diameter of a dry peptide resin bead on which a peptide with a MW of 5957 was assembled increased from 40 to 94 µm. Thus the relative volume increase was fivefold. There was no indication that the resin matrix became filled with a peptide, even at a peptide/resin weight ratio of 4:1, or that the efficiency of the synthesis became limited by increased steric hindrance. In CH_2Cl_2 the swelling ratio was 1.84 for unloaded resin, gradually decreasing as the peptide content increased. It became constant at 1.33 after the peptide content reached 2.5 g per gram of resin. In DMF the ratio was 1.49 for unloaded resin and increased to 1.84 for the same peptide-loaded resin.

In SPPS, many peptides are prepared with inadequate yields, with unknown causes. Some of the problems include incomplete coupling and deprotection steps, the formation of truncated peptides, and the deletion sequences. To gain further understanding of these synthetic problems, Marshall and co-workers (Hancock et al. 1973) investigated the source of truncated and deletion sequences in SPPS and found that dynamic solvation was a major factor. In another example, MAS NMR characterization of the SPPS product of the difficult sequence formed by the 10-residue peptide NH_2-Thr-Glu-Gly-Glu-Ala-Arg-Gly-Ser-Val-Ile-OH (Dhalluin et al. 1997) showed that resin swelling influenced the outcome of the reaction at critical synthetic steps. A detailed secondary structure determination of the growing peptide on the resin beads, based on NOE analysis and the ^1H and ^{13}C chemical shift deviations, indicated an extended structure on the whole length of the sequence for peptides with fewer than 8 residues (Figure 2.9). For 9- and 10-residue peptide resins, spectral broadening indicated an aggregation of the peptide. The addition of H-bond-accepting DMSO disrupted the peptide aggregation caused by interpeptide chain H-bonding, improved the ^1H NMR spectral line width, and increased reaction yield (36% to 61% for 10-residue peptides and 58% to 67% for 9-residue peptides) at the same time (Figure 2.10). This is because the disruption of peptide aggregates allowed the remaining residual free amino functions to be more accessible to the deprotecting and acylating reagents, resulting in better coupling efficiency at critical synthetic steps.

According to the Merrifield strategy, SPPS begins in the nonpolar environment of the PS matrix, with the environment evolving gradually toward a more polar nature as the peptide chain grows. If the nature of the solvent does not evolve in a parallel way, aggregation

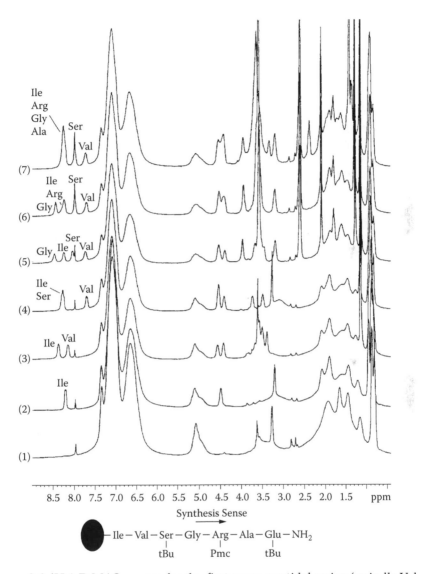

Figure 2.9 ¹H 1 D MAS spectra for the first seven peptidyl resins (resin-lle-Val-Ser(Ot-Bu)-Gly-Arg(Pmc)-Ala-Glu(Ot-Bu)-NH₂) swollen in DMF-d₇ at room temperature and 4 kHz spinning rate. The assignment of the NH resonances at each synthesis step is shown. (Reproduced with permission from *J. Am. Chem. Soc.* 1997, 119, 10494–10500. Copyright© 1997 American Chemical Society.)

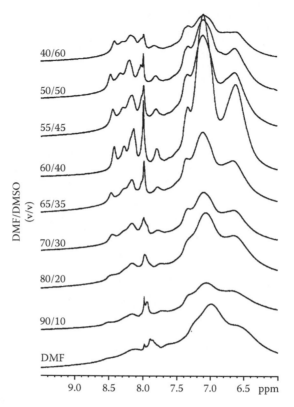

DMF/DMSO (v/v)

40/60

50/50

55/45

60/40

65/35

70/30

80/20

90/10

DMF

9.0   8.5   8.0   7.5   7.0   6.5 ppm

Figure 2.10 ¹H 1 D MAS (4kHz) spectra of the 10-residue peptidyl resin (resin-Ile-Val-Ser(Ot-Bu)-Gly-Arg(Pmc)-Ala-Glu(Ot-Bu)-Thr(Ot-Bu)-NH₂) swollen in various solvent mixtures of DMF-d₇/DMSO-d₆. (Reproduced with permission from *J. Am. Chem. Soc.* 1997, 119, 10494-10500. Copyright© 1997 American Chemical Society.)

and cross-linking of the peptide complex (depending on the sequence, the side chain protection groups, and the resin loading) may lead to an incomplete solvation of the peptide resin, shrinkage of the resin, reduced reagent penetration, and finally a decrease in coupling yield. This was further demonstrated by studying resin-linked peptide sequences using electron paramagnetic resonance (EPR) spectroscopy (Cilli et al. 1999). The swelling of resins and of peptide-bound resins is correlated with solvents, while the reaction yield depends on the swelling. The solvent preference of resin as it is swelling changes with the changing content on the resin.

In summary, chain mobility is essential for optimal yields and may be enhanced by the use of polar solvent to swell the peptide resin; and DMSO, when added to DMF, may be used to disrupt peptide aggregation. These studies of SPPS are relevant to SPOS where dynamic solvation problems need to be considered in combinatorial or parallel synthesis.

2.4 Effects of the solvated resin on solid-phase reactions

The resin support has significant effects on the physical and chemical nature of the attached reactive groups and entrapped reagent molecules. These effects can be explained as follows:

- the effects of solvated resin on SPPS
- resin support as another "solvent phase"
- support effects on SPOS reaction kinetics
- site separation and site interaction
- support effects on product purity

2.4.1 The effects of support solvation on SPPS

The influence of the chain length on the coupling reaction in SPPS was investigated by Hagenmaier (1970) for the synthesis of the dodecapeptide (Leu-Ala)$_6$ on the Merrifield resin (2% DBV, 200–400 mesh). The method of Dorman (1969) was used to determine the efficiency of the coupling reaction. The first six steps of synthesis gave almost quantitative yields after repetition of the coupling reaction using fresh BOC–amino acid plus dicyclohexylcarbodiimide. In later coupling steps, yield decreased steadily with increasing chain length and was more pronounced for leucine than for alanine owing to the greater steric hindrance in the case of leucine. Carrying out the coupling reaction for a third time did not increase the yield, which suggests that aggregation of the elongated peptide chains prevented reagent access, as discussed in the previous section (Dhalluin et al. 1997).

In a study comparing various solid supports for SPPS, five polymers (polystyrene-1% DVB (I), polystyrene macroporous (II), Kel-F-g-styrene (III), polyacrylamide (IV), and controlled pore glass (V)) were tested in order to evaluate the influence of the chemical and physical conditions of the support on the reaction (Albericio et al. 1989). The Fmoc-(60-67)-uteroglobin-OH-protected segment was coupled onto H-(68-70)-uteroglobin previously assembled on each of the solid supports (Diagram 2.3). Microporous polystyrene (I) and polyacrylamide (IV) appear to be the most convenient supports for segment coupling. This correlates very well with the smaller line widths at half-height observed for supports I and IV in the gel-phase ^{19}F NMR spectra in CDCl$_3$ or DMF (Figure 2.11) and in the ^{13}C NMR spectra of the same materials observed in CDCl$_3$. The faster coupling rates observed using (I) (2 h, 31% yield and 7 h, 100%) as compared with IV (2 h, 24% and 7 h, 87%) are in agreement with the T_1 measurements (I: 0.95 $CDCl_3$, 0.71 DMF; IV: 0.61 $CDCl_3$, 0.50 DMF) that show a clear difference between resin I and the other resins.

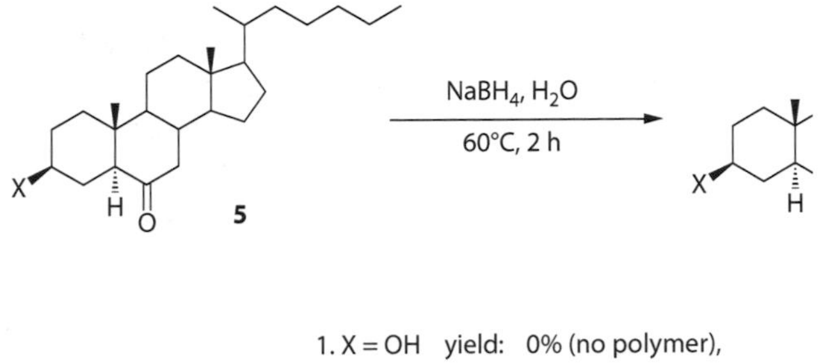

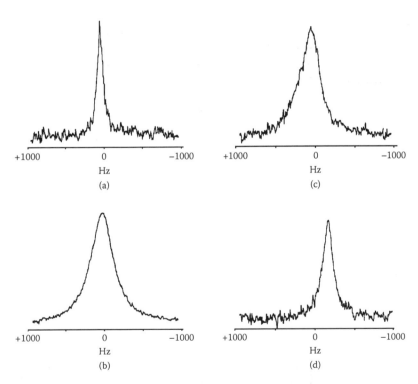

NaBH$_4$, H$_2$O

60°C, 2 h

5

1. X = OH   yield: 0% (no polymer),

Diagram 2.3

+1000   0   −1000

Hz

(a)

+1000   0   −1000

Hz

(c)

+1000   0   −1000

Hz

(b)

+1000   0   −1000

Hz

(d)

Figure 2.11 ¹⁹F NMR signal of Boc-5-flurotryptophon (BFT) bound to support I(a), II(b), III(c), and IV(d) swollen in CDCl₃. Spectra were run at 1888 MHz, with a 30-kHz spectral width using 90°C pulses and an 8-s recycling time. (Reproduced with permission from *J. Org. Chem.* 1989, 54, 360–366. Copyright© 1989 American Chemical Society.)

2.4.2 Resin support as another "solvent phase"

Evidence has accumulated that is starting to reveal the nature of the resin support as another solvent phase.

2.4.2.1 Selective adsorption of reactant and reagent molecules

A polymer-supported reaction occurs when the substrate enters the resin. One factor determining reactivity is the polarity of the resin. On an anion-substituted macroreticular resin, the selective absorption of the polar substrate into the resin increased the rate of reactions involving polar reactants (Cainelli et al. 1982).

Polymer supports can selectively adsorb reactants and create a more concentrated reactant pool inside the resin bead. For the reduction reaction shown in Scheme 2.1, both the starting material and the reagent were adsorbed into a polymer support, so that the reaction yield and the stereochemical selectivity were greatly improved compared to control experiments with no polymer present. The effect was further enhanced when a catalytic amount of tetra-*n*-butylammonium bromide (TBAB) was added. Because of the structural differences between the gel-type PS resins and the macroporous PS resins, only the latter improved the stereochemistry (Briggs and Hodge 1988).

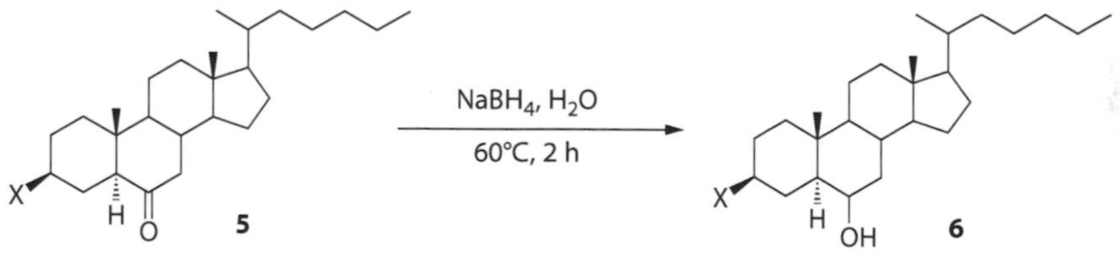

$$NaBH_4, H_2O$$
$$60°C, 2 h$$

1. X = OH   yield: 0% (no polymer),
                              65% (2% DVB anionic polymer)

2. X = H   63% (no polymer),
                              82% (2% DVB anionic polymer)

Scheme 2.1

Regen and Dulak (1977) investigated the cosolvent-like property of PS-PEG resins. Illustrated in Scheme 2.2, the polar PS-PEG resin significantly enhanced the rates and yields of these aqueous reactions. Two possible mechanisms for this resin effect are (1) the polymer may provide a common phase where a high effective concentration of the reactants is possible; (2) the microenvironment within the resin phase may influence the starting material and transition state stability.

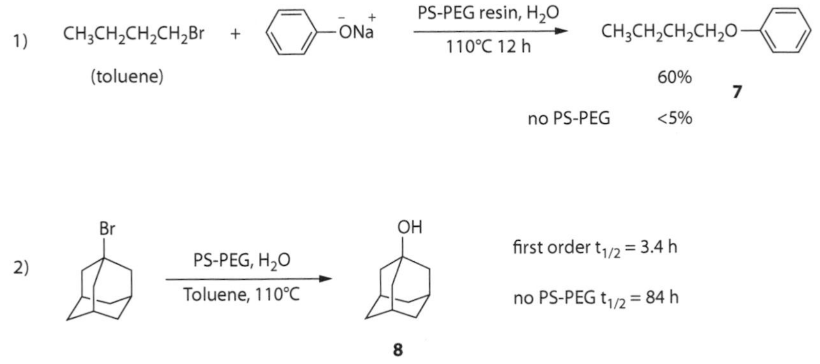

1) $CH_3CH_2CH_2CH_2Br$ + ⟨⟩—ONa⁻⁺ →[PS-PEG resin, H₂O / 110°C 12 h] $CH_3CH_2CH_2CH_2O$—⟨⟩

(toluene)   60% **7**

no PS-PEG   <5%

2) [Br-adamantane structure] →[PS-PEG, H₂O / Toluene, 110°C] [OH-adamantane structure]   first order $t_{1/2} = 3.4$ h

no PS-PEG $t_{1/2} = 84$ h

8

Scheme 2.2

2.4.2.2   The effect of phase-transfer catalyst (PTC) on SPOS reactions

Without using PTC, the reactions shown in Scheme 2.3 gave 30% yield; with PTC, yields of 88–98% were achieved (Frechet et al. 1979a). PTCs have also facilitated a range of S_N2 reactions on the Merrifield resin using amines, acids, salt, alcohol, and phenols (N'Guyen and Boileau 1979). Three series of aqueous SPOS reactions (nucleophilic substitutions on the Merrifield resin; nucleophilic addition of CN⁻ to formylpolystyrene resin; and reactions involving polymeric nucleophiles) were studied by Frechet et al. (1979b). These results demonstrated remarkable success for phase-transfer cataly-sis and indicated that the swollen resin can be viewed as another "solvent phase" in terms of its effects on organic substrates and reactions.

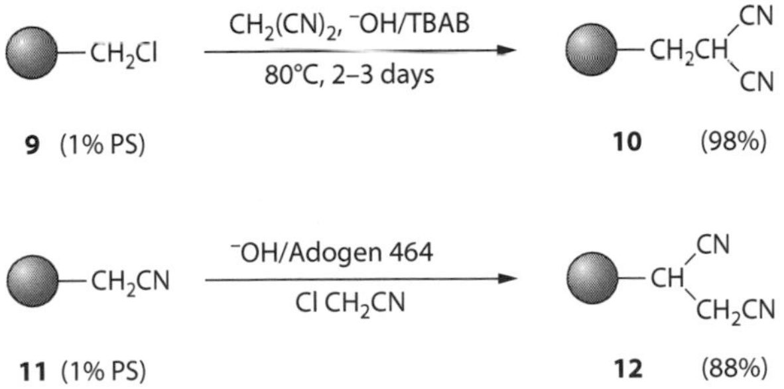

●—CH_2Cl →[$CH_2(CN)_2$, ⁻OH/TBAB / 80°C, 2–3 days] ●—CH_2CH⟨$\substack{CN \\ CN}$

9 (1% PS)   **10**   (98%)

●—CH_2CN →[⁻OH/Adogen 464 / Cl CH₂CN] ●—CH⟨$\substack{CN \\ CH_2CN}$

11 (1% PS)   **12**   (88%)

Scheme 2.3

2.4.3   Support effects on SPOS reaction kinetics

The stereotypical view of a resin bead is a rigid, non-solution-like matrix into which organic reagents diffuse after having to overcome a steric bar-rier. Although the resins are usually heavily solvated (swollen resin beads contain 80% solvent), reactions are undoubtedly affected by the proximity of the attached reactant to the spacer and the polymer matrix. Synthesis using soluble polyethyleneglycol (PEG) polymers was often considered in order to take advantages of certain benefits of both solution- and solid-phase synthesis. An alternative approach has been the development of PS-PEG resins designed to combine useful features from both insoluble PS support and the soluble PEG support.

Kinetics of a series of organic reactions (Scheme 2.4) on both PS and PS-PEG resins have been determined using single-bead FTIR microspectroscopy (Li and Yan 1998). The results are listed in Table 2.1. It is generally assumed that reactions proceed faster on PS-PEG resin because it provides a more "solution-like" matrix compared with PS resins. The so-called "solution-like" matrix means that the attached reactant is floating in a mobile PEG phase within the PS-PEG resin instead of being attached to an "immobile" PS phase. This hypothesis may explain some results of solid-phase peptide synthesis and some organic reactions involving polar reagents. For example, oxidation of an alcohol (reaction 1 in Scheme 2.4) is more rapid on PS-PEG than on PS resin. However, of the four reaction types studied in this investigation, two reactions exhibited higher reaction rates when carried out on PS resins than on PS-PEG resins.

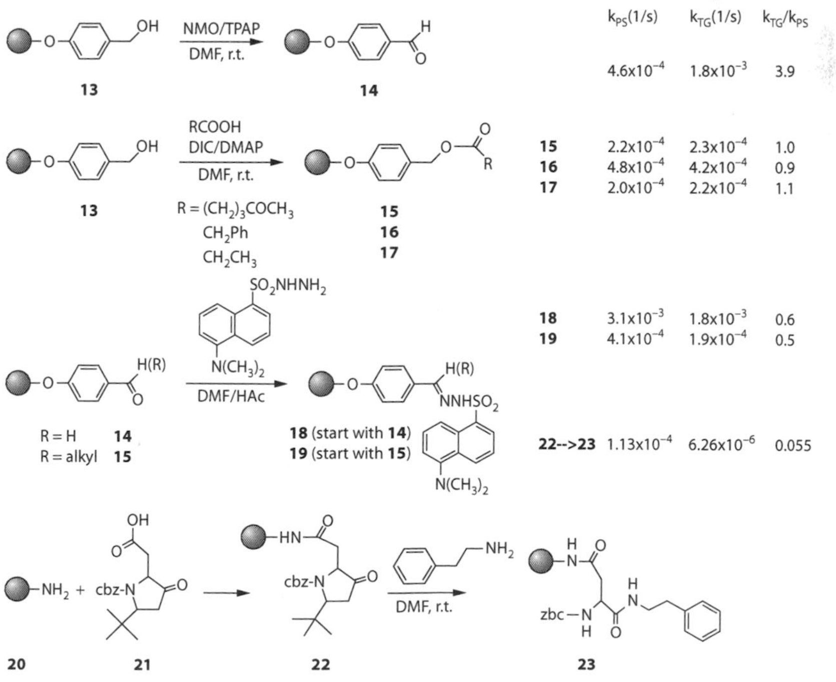

Scheme 2.4

Table 2.1 Comparison of Reaction Rates on PS- and
TG-Based Resins

Reaction scheme	k_{PS} (1/s)	k_{TG} (1/s)	k_{TG}/k_{PS}
1	4.6×10^{-4}	1.8×10^{-3}	3.9
2 (R = 3)	2.2×10^{-4}	2.3×10^{-4}	1.0
2 (R = 4)	4.8×10^{-4}	4.2×10^{-4}	0.9
2 (R = 5)	2.0×10^{-4}	2.2×10^{-4}	1.1
3 (R = H)	3.1×10^{-3}	1.8×10^{-3}	0.6
3 (R = 3)	4.1×10^{-4}	1.9×10^{-4}	0.5
4 (step 2)	1.13×10^{-4}	6.26×10^{-6}	0.055

ESR studies have shown that the rate of movement of the pendant group is high in swollen PS resin. Merrifield pointed out (1984) that this extensive chain flexibility plays an important role in the rate and extent of the reactions in solid-phase peptide synthesis. In previous kinetics studies (Yan et al. 1996a, 1996b; Sun and Yan 1998; Li and Yan 1998), it was found that, except for trityl-like linkers, the proximity of the reactive group to the spacer and the PS backbones in general poses little steric hindrance. This may be because full solvation and the dynamic fluctuation of the PS matrix allows free access of the reagent into the resin bead. The steric hindrance assumption cannot explain the results in Table 2.1. What then determines the relative reaction rate on different resin supports?

Czarnik (1998) summarized some key observations on this subject and pointed out that solid-phase synthesis supports are like solvents. The effect of the polymer matrix on the reaction rate may be very similar to the effect of the solvent on the rate of a solution reaction. A swollen bead is the ultimate microreactor for SPOS. Although 80% of the bead is solvent, spacers and matrix exert significant effects on incoming reagents and reactants depending on their molecular properties, such as polarity. The solvent-like support effect and the effect of phase-transfer catalysts on the solid-phase reaction discussed in the previous section help substantiate this hypothesis. A complete understanding of how the structure of a support affects the reactivity of an attached reactant awaits further investigation.

2.4.4 Site separation and site interaction

Although solvent swelling can improve resin accessibility to reagents, it also creates some unwanted effects, such as site–site interactions. Intuitively, the site–site distance should increase with swelling of the resin, but actually the dynamic distance between reactive sites becomes smaller, such that site–site interactions occur. Pendant moieties on PS resins do not maintain the separation implied by percent substitution and the well-defined bond lengths in polystyrene. The expected site separation and "pseudo-dilution" do not hold after resin swelling. Because of the long PEG spacers, site–site interaction is much more extensive in PS-PEG resins. Since the interactions between reactive groups inside the resin can alter the SPOS reactions, it is critical to fully understand the site separation and site-interaction effects in PS resins (Shi et al. 2007). In this section, the sequence of discussion will be

- site separation effect
- site interaction effect
- factors affecting site separation/interaction

2.4.4.1   Site-separation effect

When resin loading is low, sites can be effectively separated. This can be seen in EPR experiments (Marchetto et al. 2005). Figure 2.12 shows that at low-loading, a sharp EPR spectrum is observed, indicating an isolated EPR probe. Good swelling or less viscous solvents can increase the mobility of the sites and promote site–site contact, thereby broadening the EPR signal. At higher loading, signals are broader, indicating significant site–site interactions.

Cyclic peptides are usually prepared from linear peptides by intramolecular cyclization reactions. The carboxylic end of the peptide is usually activated by the formation of an active ester, anhydrides, azides, or acid chlorides, with the free terminal amino end being allowed to react

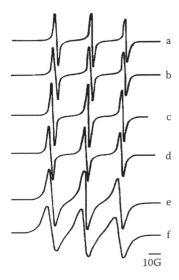

Figure 2.12 Effect of Boc-TOAC-OH loading on the EPR spectra of Boc-TOAC-BHAR in DCM (a) and DMF (b). Probe loading (mmol/g): a = 0.003; b = 0.019; c = 0.035; d = 0.050; e = 0.065; f = 0.134; g = 0.646; h = 0.988. (Reproduced with permission from *J. Org. Chem.* 2005, 70, 4561–4568. Copyright© 2005 American Chemical Society.)

with the active terminal carbonyl group at high dilution. As the result of intermolecular condensation occurring under these conditions, linear oligopeptides are formed in addition to the desired cyclic peptide, so that solution-phase synthesis results in reaction mixtures from which cyclic peptides are isolated usually in relatively low yields.

Fridkin et al. (1965) used two polymer reagents in the preparation of the peptide esters: cross-linked poly-4-hydroxy-3-nitrostyrene, and a branched copolymer of DL-lysine and 3-nitro-L-tyrosine. Cyclo(Gly-Gly) was synthesized on these two polymers in 75% yield (DMF, r.t. 12 h). Cyclo(tetra-L-Ala) was also synthesized on the polymer (r.t. 12 h) with yields of chromatographically pure peptide of 50–60% compared with 40% obtained from solution-phase synthesis from tetra-L-Ala ester (pyridine, 100°C, 12 h).

Flanigan and Marshall (1970) also reported solid-phase synthesis of cyclic peptides at higher overall yields with a minimization of racemization during activation and subsequent cyclization. Five cyclic peptides were synthesized on polymer supports with yields of 40–60%.

Camps et al. (1971) observed hydroxyl band splitting (3580 and 3460 cm^{-1}) and pointed out that in 2% DVB PS, the hydroxyl groups might be divided into free form at 3580 cm^{-1} and the associated (hydrogen-bonded) hydroxyl groups at 3460 cm^{-1} (Yan and Sun 1998). The experimental proof was obtained later by demonstrating that the addition of the hydrogen-bond-accepting solvent DMSO converted the free hydroxyl band at 3580 cm^{-1} to the hydrogen-bonded band at 3460 cm^{-1}. Alternatively, increased steric hindrance around the hydroxyl group converted the band at 3460 cm^{-1} to the free hydroxyl band at 3580 cm^{-1} (Yan and Sun 1998).

Carboxyl resin exhibits two carbonyl bands in the IR spectrum: a strong band at 1694 cm^{-1} and a weak band at 1727 cm^{-1}. Letsinger et al. (1964) pointed out that the former may be attributable to the hydrogen-bonded carboxyl groups and the latter to the free carboxyl groups.

The most important factor governing intrapolymeric reaction is the frequency of encounters between reactive substituents. Unlike the diffusion-limited rate of collisions between monomeric reagents in solution, the encounter rate of reactive groups in a polymer is limited by the potentially more restricted conformational motions of the polymer backbone. Mazur and Jayalekshmy (1976, 1979) carried out a crude measurement of the encounter frequency of substituents focusing on the chemistry of *o*-benzyne (1,2-dehydrobenzene) covalently bound to a PS support (Scheme 2.5). While monomeric benzyne undergoes very rapid dimerization in solution, this pathway was found to be greatly retarded for the polymer-bound analog. *o*-Benzyne was first generated by oxidation of 1-aminobenzotriazole derivatives bound to various PS supports. Upon oxidization, the benzyne intermediate was converted to a mixture of isomeric aryl acetates. This reaction was not observed for monomeric 1-aminobenzotriazole, where dimerization was the predominant pathway.

The frequency of encounters between the polymer-bound benzyne intermediates allows the relatively slow acetate formation to compete with the encounter-controlled dimerization process. Delayed trapping of the benzyne intermediate by a diene through a Diels-Alder reaction demonstrated that the frequency of encounter between the intermediates bound to a 2% cross-linked PS resin must lie between 2.6×10^{-2} and 1.1×10^{-3}. This prolonged lifetime for the resin-bound benzyne molecules indicated considerable site separation within the bead.

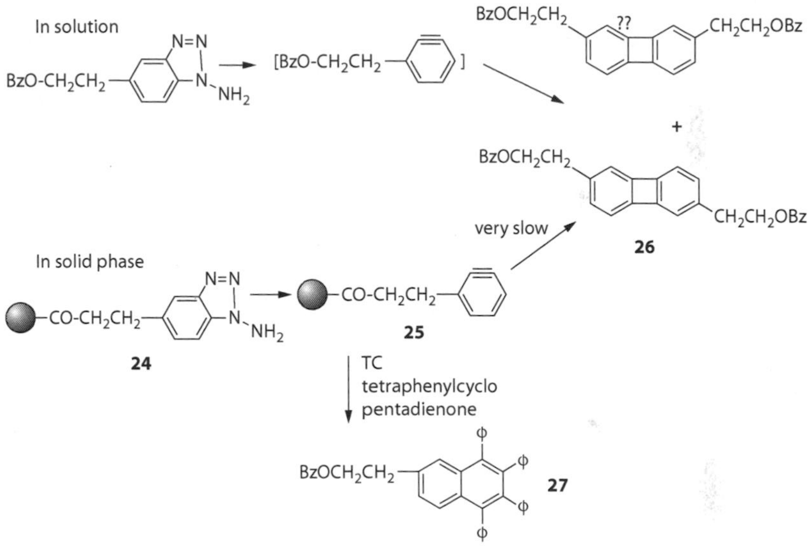

Scheme 2.5

Scheme 2.6 shows an intrapolymeric glycyl transfer, or an intraresin nucleophilic reaction (Rebek Jr. and Trend 1979). The empirical value of 1–2 min for the half-life of this compound, assuming a second-order process for its disappearance, was derived from an acid-anhydride quenching study. If acyl transfers between the sites of the compound do not occur at every encounter, this half-life corresponds to a lower limit for the encounter frequencies of reagents attached to such polymers. This figure agrees well with that derived by Mazur and Jayalekshmy for the dimerization rate of polymer-bound benzyne (1976, 1979).

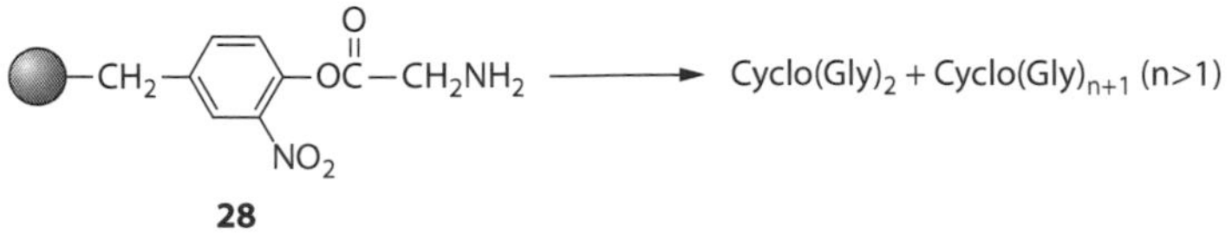

Scheme 2.6

Cyclopentadienyl iron (II) and cobalt (I) carbonyl complexes were attached to 18% DVB-PS resin. Because of the carbonyl stretching vibration in the IR spectrum, the iron complex remains unchanged for months at room temperature under the exclusion of light and air. In contrast, solution-phase $C_5H_5Fe(CO)_2H$ has been observed to be a highly unstable species, which rapidly loses H_2 to form binuclear $(C_5H_5)_2Fe_2(CO)_4$. The absence of any μ-CO absorption, typical of such a binuclear particle, from the polymer–supported iron compound substantiates the view that attachment to a cross-linked polymer can suppress any metal–metal interactions and that the unusual stability of the polymer-supported iron hydride species is a consequence of their mutual isolation by the polymer network (Gubitosa et al. 1977).

Similarly, photolysis of $C_5H_5Co(CO)_2$ in a solution leads to a succession of di- and tri-nuclear clusters such as $(C_5H_5)_2Co_2(CO)_2$, $(C_5H_5)_2Co_2(CO)_3$ and $(C_5H_5)_3Co_3(CO)_3$ characterized by IR absorption around 1970, 1800, and 1670 cm^{-1}. However, when polymer-bound $C_5H_5Co(CO)_2$ is irradiated in a petroleum ether (a poor swelling solvent) suspension at an ambient temperature with an unfiltered medium-pressure Hg or Xe lamp, a disappearance of the absorption at 1960 and 2020 cm^{-1} is observed, indicating the photodissociation of the CO ligands. No trace of any μ-CO band is observed at wavelengths below 1900 cm^{-1}. These data clearly indicate the absence of any CO-bridged, di-, or polynuclear species (Gubitosa et al. 1977).

Taking advantage of the site–site separation effect, Leznoff has performed a series of successful monofunctionalization reactions of diols and dialdehyde (Leznoff and Wong 1972; Wong and Leznoff 1973a; Wong and Leznoff 1973b).

2.4.4.2   Site-interaction effect

Titanocene with various loading was attached to 20% DVB PS resin, and the hydrogenation of the 1-hexene was studied by polymer-supported Cp_2TiCl_2 at room temperature. The rate of hydrogenation in mLmin^{-1} mmol^{-1} of Ti indicated the percentage of the titanium centers that were active. As the loading increased, the percentage of metal centers close enough to interact increased, and consequently the number of potential active sites eventually decreased through site–site interaction and destruction. The rate of hydrogenation of 1-hexene as a function of the loading of Ti on the polymer beads was determined, and the result was a bell-shaped curve with a maximum (Figure 2.13) (Grubbs et al. 1977). This indicated that the rate of catalysis increased as the Ti loading increased, until site–site interaction caused a loss of active sites.

A relationship between resin-bound mercapto groups has been studied (Wulff and Schulze 1978). Since mercapto groups can be selectively re-oxidized to disulfides (Scheme 2.11), their spatial relationship can be quantitatively determined on the basis of the re-oxidized portion. The re-oxidation of the SH groups was carried out with iodine in methanol. Polymers with less cross-linking exhibited higher degrees of re-oxidation

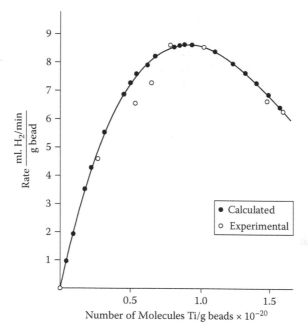

Figure 2.13 Rate of hydrogenation of 10-hexene as a function of the loading of Ti on the polymer beads. (Reproduced with permission from *J. Am. Chem. Soc.* 1977, 119, 4517–4518. Copyright© 1977 American Chemical Society.)

(69%). Remarkably, a 33% re-oxidation has been observed even in highly cross-linked polymers, suggesting that the polymer chains retain considerable flexibility, making reactions possible between widely distributed SH groups. A higher degree of oxidation of mercapto groups in resin swollen in benzene (a good swelling solvent for PS resins) indicated that the increased mobility of polymer chains in better-swollen resin promotes site interaction (Wulff and Schulze 1978).

Crowley and Rapoport (1970) carried out a series of studies of Dieckmann cyclization reactions. In the Dieckmann cyclization (Crowley et al. 1973) shown in Scheme 2.7, the cyclized product was formed in 19% yield. This dimerization occurred undoubtedly because the individual moieties were not effectively separated from each other during the reaction. The best interpretation for this site–site interaction is the resin flexibility during the reaction.

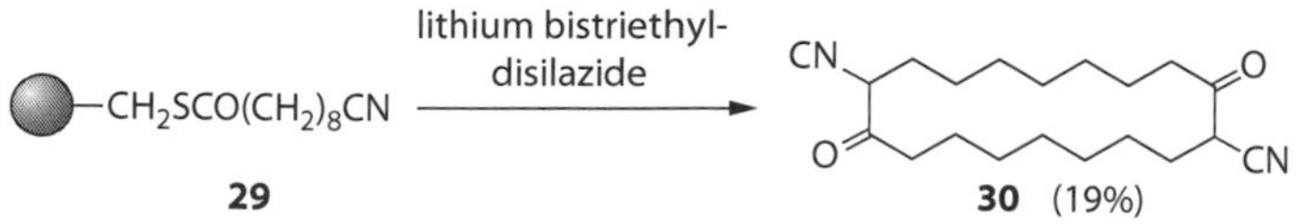

Scheme 2.7

The utilization of a polymeric phenylthiomethyllithium reagent for the homologation of alkyl iodides has been reported (Crosby and Kato 1977). The use of 1,4-diiodobutane provided an appropriate system for the study of intraresin reactions of polymer-bound ionic functional groups. The ratio of 1,5- to 1,6-diiodoalkanes obtained by the two-step (Scheme 2.8) homologation sequence was a direct indication of the interaction of reactive sites.

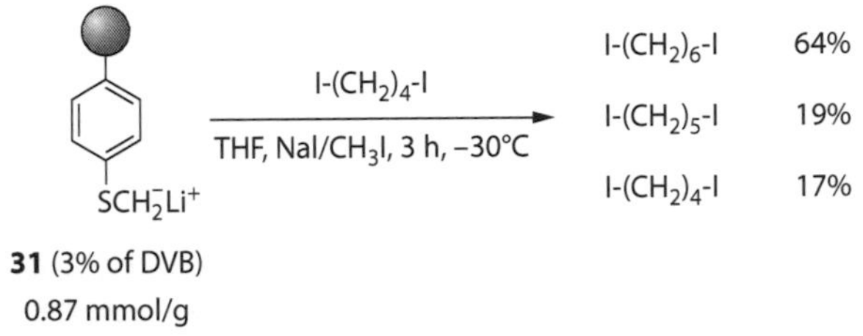
I-(CH2)4-I
THF, NaI/CH3I, 3 h, –30°C
I-(CH2)6-I 64%
I-(CH2)5-I 19%
I-(CH2)4-I 17%
SCH2Li+
31 (3% of DVB)
0.87 mmol/g

Scheme 2.8

To take advantage of the site–site interaction effect, a mixed condensation product was synthesized by purposefully attaching two different kinds of reactants onto resin (Scheme 2.9) (Kraus and Patchornik 1971).

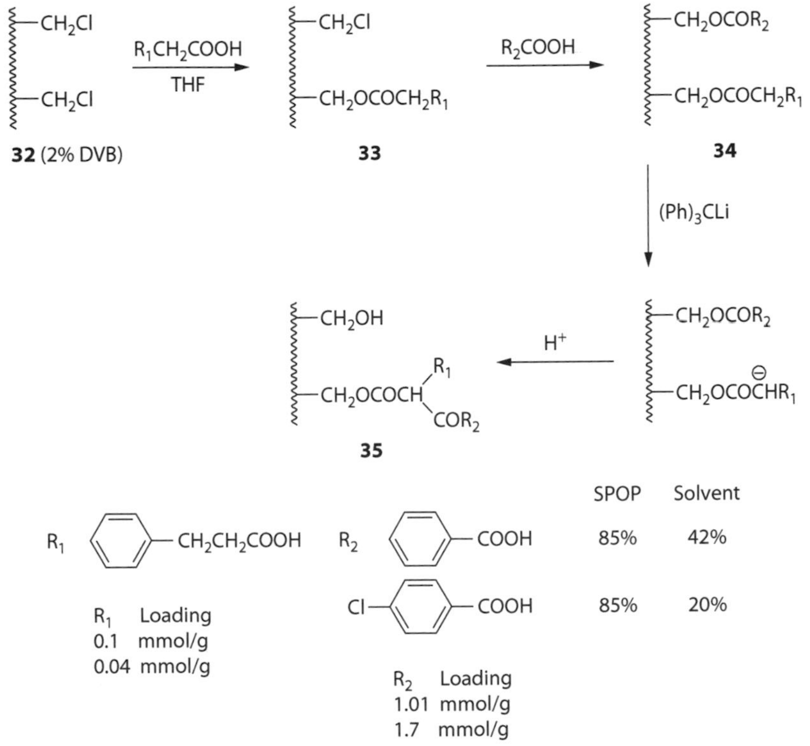
CH2Cl
CH2Cl
R1CH2COOH
THF
32 (2% DVB)
CH2OCOCH2R1
33
R2COOH
CH2OCOR2
34
(Ph)3CLi
CH2OCOCHR1
H+
CH2OH
CH2OCOCH
R1
COR2
35
SPOP Solvent
R1 CH2CH2COOH R2 COOH 85% 42%
Cl COOH 85% 20%
R1 Loading
0.1 mmol/g
0.04 mmol/g
R2 Loading
1.01 mmol/g
1.7 mmol/g

Scheme 2.9

2.4.4.3   Factors affecting site separation

The equilibrium between site–site separation and site-site interaction can be explained by favorable kinetics relationships between the desired reactions and competitive reactions. When the less kinetically favorable reactions are attempted with polymer-bound reagents, the extent of site–site interactions will increase.

Resin swelling greatly increases site–site interactions. Two decomposition pathways of resin-bound carbonic benzoic anhydride have been investigated under various conditions. The intrastrand reaction resulted in the formation of the benzoate ester on the resin (A), while the interstrand resulted in the formation of the benzoic anhydride and polymeric carbonate (B) (Scheme 2.10). The latter pathway is favored in solvent-swollen resin, and the former in dry resin (Martin et al. 1978). A higher degree of oxidation of mercapto groups in resin swollen in benzene indicated the better mobility of polymer chains in better-swollen resin promotes site interactions, as shown above (Wulff and Schulze 1978).

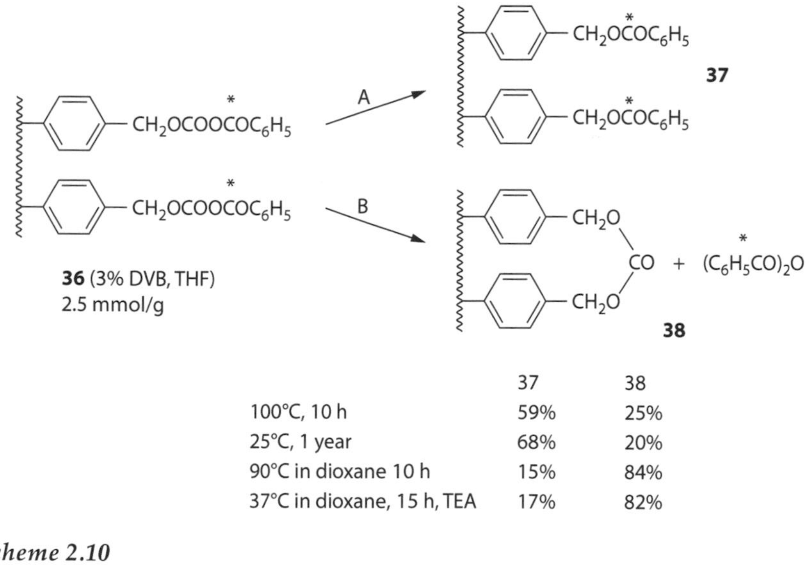

	37	38
100°C, 10 h	59%	25%
25°C, 1 year	68%	20%
90°C in dioxane 10 h	15%	84%
37°C in dioxane, 15 h, TEA	17%	82%

Scheme 2.10

Regen and Lee (1977) used a method introduced by Collman (1972), which consists of IR measurements of the 1:1 and 2:1 complex formed between 2% DVB cross-linked styryl diphenylphosphine and Co(NO)(CO)$_3$. A high degree of site isolation promotes the formation of the 1:1 complex (IR band of NO at 1755 cm^{-1}), and a high degree of site interaction promotes the formation of the 2:1 complex (IR band at 1710 cm^{-1}) between the resin-bound phosphine and Co(NO)(CO)$_3$. Starting with the 1:1 complex, IR indicated that the addition of very small quantities of the good-swelling solvent m-xylene to the resin resulted in a rapid increase in site–site interaction and the formation of the 2:1 complex at the expense of the 1:1 complex (Figure 2.14). In the absence of solvent or in the presence of an excess of a poor-swelling solvent such as n-hexadecane, no change was observed in the IR spectrum from the starting 1:1 complex. These results clearly establish that bead swelling is an important factor in site isolation in low cross-linked PS resins.

It is interesting to compare results from different laboratories on this topic. For example, Grubbs and Kroll (1971) reported observation of site isolation on a macroreticular 20% cross-linked PS resin using the poor-swelling solvent cyclohexane. In contrast, Scott et al. (1977) found significant site–site interaction in the same resin when the good-swelling solvent CH_2Cl_2 was used. These results indicated the influence of a solvent in determining site isolation on highly cross-linked resins. The intensity of the free hydroxyl band at 3580 cm⁻¹ increases as the microenvironment of the hydroxyl group becomes bulkier. By controlling both the steric hindrance and the local concentration, total site isolation can be achieved, as shown in Figure 2.15.

Site–site interaction can also be controlled by altering the reacting group/reactive site ratio. In the reaction of cross-linked chloromethyl PS with 1,4-butanedithiol, the site interaction was controlled by using a different ratio of the starting materials (Scheme 2.11) (Wulff and Schulze 1978). The interstrand reacting product was reduced from 97% to 26% when the excess of dithiol was raised from 3 fold to 11 fold. The reaction of Wang resin with a series of diacid chloride (Scheme 2.12) showed both the

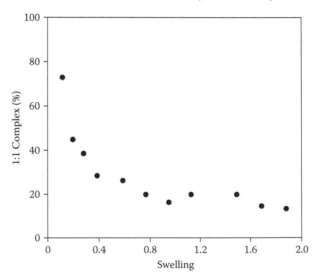

Figure 2.14 Plot of percent of 1:1 complex remaining as a function of degree of swelling, g of m-xylene/g of dry resin, of the resin after a reaction time of 240 h at 70°C as estimated through relative IR band intensities. (Reproduced with permission from *Macromolecules* 1977 10, 1418–1419. Copyright© 1977 American Chemical Society.)

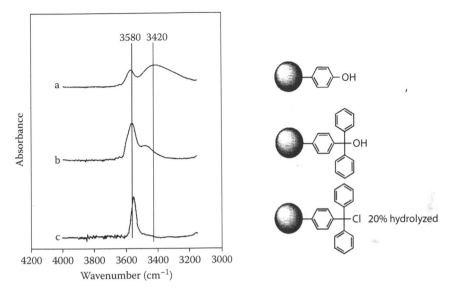

Figure 2.15 IR spectra taken from a single flattened bead for resins shown on the right. (a) Wang resin; (b) tritylhydroxyl resin; (c) tritylchloride resin with ~20% hydrolysis (to tritylhydroxyl resin).

dynamic nature of the resin (lack of diacid chloride length dependence) and the effect of the molecular ratio. The interstrand product was reduced from 72% to 54% when the excess of diacid chloride was raised from 2 fold to 10 fold (Yan and Sun 1998).

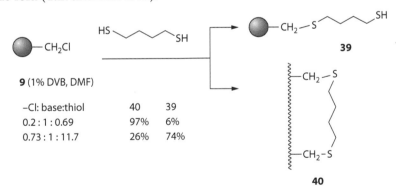

–Cl: base:thiol	40	39
0.2 : 1 : 0.69	97%	6%
0.73 : 1 : 11.7	26%	74%

Scheme 2.11

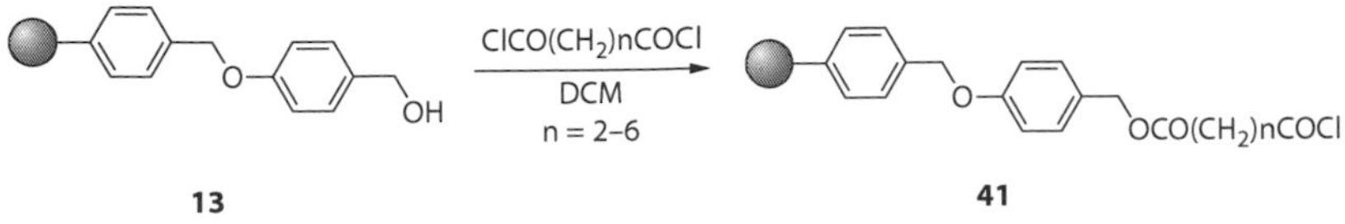

Scheme 2.12

2.4.5   Solid support effects on product purity

Because the SPOS resin itself is a synthetic product, the quality of resins from different commercial sources or different batches from the
same source may vary. Some resins may contain residues of covalently
or noncovalently bound synthetic intermediates (such as carboxylpolystyrene resin); others may have undergone partial hydrolysis during
handling (such as tritylchloride resin) or may have changed their loading value during storage (such as formylpolystyrene). The result of any of
these problems could be impurities in the final cleaved synthetic product.
Since it is often impractical to purify combinatorial libraries or products
from parallel synthesis, the impact of the resin on the quality of the synthetic product is immense. In order to minimize some of these problems,
MacDonald et al. (1995) used prewashing to eliminate entrapped resin
impurities, and this approach is recommended for the routine removal
of noncovalently bound impurities. They also suggested the use of mild
cleavage conditions to liberate the desired compounds without causing
resin breakdown or impurity cleavage.

2.5   Summary

SPOS is highly sensitive to the microenvironment of the resin and other
reaction conditions. When designing the synthesis of a combinatorial
library, the influence of the resin after solvation and swelling must be
carefully considered. This assures the quality of the final library that is
often too large to be completely characterized. The physicochemical and
swelling properties of resin supports have direct impact on many spectroscopic analysis methods, such as NMR.

Selective solvation of resin alters the local reactivity and accessibility
of the bound substrate as well as the mobility of the entrapped reagent.
Resin solvation changes during the course of the reaction if the attached
substrate changes the polarity or other physical and chemical properties.
In addition, swollen beads can selectively adsorb compounds and thereby
influence the reaction kinetics. The use of a phase-transfer catalyst can
enhance the rate of reaction on resin and illustrates the solvent-like property of resin supports. Swollen resin can allow sites to interact with one
another in 1% DVB resins in varying degrees depending on solvent, resin,
and reactivity of the pendant groups, so that total site isolation seems
achievable only by controlling several factors simultaneously, such as by
lowering loading, increasing steric hindrance, and increasing the degree
of cross-linking.

The control of these properties during SPOS for combinatorial chemistry involves the proper selection of resin and solvent, the alternation of
solvents to accommodate dynamic solvation of the resin, the optimization

of kinetics when changing solid supports, and a careful control of resin impurities. This will lead to a synthesis of a high-quality combinatorial library. The study of solid-support effects on SPOS is an area that needs more attention. Also needed is a clarification of the influence of high and low temperature effects, resin compatibility with various organic molecules, and dynamic solvation in SPOS.

References

Albericio, F., Pons, M., Pedroso, E., Giralt, E. 1989. Comparative study of supports for solid-phase coupling of protected-peptide segments. *J. Org. Chem.* 54, 360.

Atherton, A., Brown, E., Sheppard, R.C. 1981. A physically supported gel polymer for low pressure, continuous flow solid phase reactions: Application to solid phase peptide synthesis. *J. Chem. Soc. Chem. Commun.* 1151.

Bray, A.M., Maeji, N.J., Geysen, H.M. 1990. The simultaneous multiple production of solution phase peptides: Assessment of the Geysen method of simultaneous peptide synthesis. *Tetra. Lett.* 31, 5811.

Briggs, J.C., Hodge, P. 1988. Potassium borohydride reductions of ketones absorbed into polymer supports: Stereochemical effects. *J. Chem. Soc. Chem. Commun.* 310.

Cainelli, G., Contento, M., Manoscalchi, F., Plessi, L. 1982. Polymer-supported reagents: Polar effects in substrate selectivity. *J. Chem. Soc. Chem. Commun.* 725.

Camps, F., Castells, J., Ferrando, M.J., Font, J. 1971. Organic syntheses with functionalized polymers: I. Preparation of polymeric substrates and alkylation of esters. *Tetra. Lett.* 20, 1713.

Cho, J.K., Park, B.D., Park, K.B., Lee, Y.S. 2002. Preparation of core-shell-type poly(ethylene flyco)-grafted polystyrene resins and their characteristics in solid-phase peptide synthesis. *Macromol. Chem. Phys.* 203, 2211.

Cilli, E.M., Oliveira, E., Marchetto, R., Nakaie, C.R. 1996. Correlation between solvation of peptide-resins and solvent properties. *J. Org. Chem.* 61, 8992.

Cilli, E.M., Marchetto, R., Schreier, S., Nakaie, C.R. 1999. Correlation between the mobility of spin-labeled peptide chains and resin salvation: An approach to optimize the synthesis of aggregating sequences. *J. Org. Chem.* 64, 9118.

Collman, J.P., Hegedus, L.S., Cooke, M.P., Norton, J.R., Dolcetti, G., Marquardt, D.N. 1972. Resin-bound transition metal complexes. *J. Am. Chem. Soc.* 94, 1789.

Crosby, G.A., Kato, M.J. 1977. The utilization of a polymeric phenylthiomethyllithium reagent for the homologation of alkyl iodides and its application for the study of intraresin reactions of polymer-bound functional groups. *J. Am. Chem. Soc.* 99, 278.

Crowley, J.I., Harvey, III., T.B., Rapoport, H. 1973. Solid phase synthesis: Evidence for and quantification of intra-resin reactions. *J. Macromol. Sci.-Chem. A.* 1117.

Crowley, J.I., Rapoport, H. 1970. Cyclization via solid phase synthesis: Unidirectional Dieckmann products from solid phase and benzyl triethylcarbinyl pimelates. *J. Am. Chem. Soc.* 92, 6363.

Czarnik, A.W. 1998. Solid-phase synthesis supports are like solvents. *Biotech. Bioeng. (Combin. Chem.)* 61, 77.

Dhalluin, C., Boutillon, C., Tartar, A., Lippens, G. 1997. Magic angle spinning nuclear magnetic resonance in solid-phase peptide synthesis. *J. Am. Chem. Soc.* 119, 10494.

Dorman, L.C. 1969. A non-destructive method for the determination of completeness of coupling reactions in solid phase peptide synthesis. *Tetra. Lett.* 2319.

Egner, B.J., Rana, S., Smith, H., Bouloc, N., Frey, J.G., Brocklesby, W.S., Bradley, M. 1997. Tagging in combinatorial chemistry: The use of coloured and fluorescent beads. *Chem. Comm.* 735.

Fields, G.B., Fields, C.G. 1991. Solvation effects in solid-phase peptide synthesis. *J. Am. Chem. Soc.* 113, 4202.

Flanigan, E., Marshall, G.R. 1970. Synthesis of cyclic peptide on dual function supports. *Tetra. Lett.* 27, 2403.

Ford, W.T., Ackerson, B.J., Blum, F.D., Periyasamy, M., Pickup, S. 1987. Self-diffusion coefficients of solvents in polystyrene gels. *J. Am. Chem. Soc.* 109, 7276.

Ford, W.T., Periyasamy, M., Spivey, H.O. 1984. Mechanisms of polymer-supported catalysis. 7. Carbon-13 NMR relaxation of toluene in cross-linked polystyrene phase-transfer catalyst gels. *Macromolecules* 17, 2881.

Forman, F.W., Sucholeiki, I. 1995. Solid-phase synthesis of biaryls via the stille reaction. *J. Org. Chem.* 60, 523.

Frechet, J.M.J., de Smet, M.D., Farrall, M.J. 1979a. Application of phase-transfer catalysis to the chemical modification of cross-linked polystyrene resins. *J. Org. Chem.* 44, 1774.

Frechet, J.M.J., de Smet, M., Farrall, M.J. 1979b. Chemical modification of cross-linked resins by phase transfer catalysis: Preparation of polymer-bound dinitriles and diamines. *Tetra. Lett.* 137.

Frechet, J.M.J., Schuerch, C. 1971. Solid-phase synthesis of oligosaccharides. I. Preparation of the solid support. Poly[p-(1-propen-3-ol-1-yl)styrene]. *J. Am. Chem. Soc.* 93, 492.

Freeman, C.E., Howard, A.G. 2001. Analysis of single polymer beads for solid phase combinatorial synthesis: The determination of reactive thiol and within batch polydispersity of bead loadings. *Analyst* 126, 538.

Fridkin, M., Patchornik, A., Katchalski, E.J. 1965. A synthesis of cyclic peptides utilizing high molecular weight carriers. *J. Am. Chem. Soc.* 87, 4646.

Gambs, C., Dickerson, T.J., Mahajan, S., Pasternack, L.B., Janda, K.D. 2003. High-resolution diffusion-ordered spectroscopy to probe the microenvironment of JandaJel and Merrifield resins. *J. Org. Chem.* 68, 3673.

Groth, T., Grotli, M., Meldal, M. 2001. Diffusion of reagents in macrobeads. *J. Comb. Chem.* 3, 461.

Grubbs, R., Lau, C.P., Cukier, R., Brubaker, Jr. C. 1977. Polymer attached metallocenes evidence for site isolation. *J. Am. Chem. Soc.* 99, 4517.

Grubbs, R.H., Kroll, L.C. 1971. Catalytic reduction of olefins with a polymer-supported Rhodium(I) catalyst. *J. Am. Chem. Soc.* 93, 3062.

Gubitosa, G., Boldt, M., Brintzinger, H.H. 1977. Supported cyclopentadienyl metal carbonyl complexes. 1. Mononuclear iron(II) and cobalt(I) to a polystyrene support. *J. Am. Chem. Soc.* 99, 5174.

Guyot, A. 1988. Synthesis and structure of polymer supports. In *Syntheses and Separations Using Functional Polymers*. Sherrington, D. C. Hodge, P. eds. 1–42.

Hagenmaier, H. 1970. The influence of the chain length on the coupling reaction in solid phase peptide synthesis. *Tetra. Lett.* 283.

Hall, P.J., Galan, D.G., Machado, W.R., Mondragon, F., Barria, E.B., Sherrington, D.C., Calo, J.M. 1997. Use of contrast-enhanced small-angle neutron scattering to monitor the effects of solvent swelling on the pore structure of styrene-divinylbenzene resins. *J. Chem. Soc., Faraday Trans.* 93, 463.

Hancock, W.S., Prescott, D.J., Vagelos, P.R., Marshall, G.R. 1973. Solvation of the polymer matrix: Source of truncated and deletion sequences in solid phase synthesis. *J. Org. Chem.* 38, 774.

Hergenrother, P.J., Depew, K.M., Schreiber, S.L. 2000. Small-molecule microarrays: Covalent attachment and screening of alcohol-containing small molecules on glass slides. *J. Am. Chem. Soc.* 122, 7849–7850.

Hodge, P. 1997. Polymer-supported organic reactions: What takes places in the beads? *Chem. Soc. Rev.* 26, 417.

Hodges, J.C., Hariskrishnan, L.S., Ault-Justus, S. 2000. Preparation of designer resins via living free radical polymerization of functional monomers on solid support. *J. Comb. Chem.* 2, 80.

Keifer, P.A. 1996. Influence of resin structure, tether length, and solvent upon the high-resolution ^{1}H NMR spectra of solid-phase-synthesis resins. *J. Org. Chem.* 61, 1558.

Kraus, M.A., Patchornik, A. 1978. Polymeric carriers as immobilizing media: Fact and fiction. *Israel J. Chem.* 17, 298.

Kraus, M.A., Patchornik, A. 1971. The directed mixed ester condensation of two acids bound to a common polymer backbone. *J. Am. Chem. Soc.* 93, 7325.

Kress, J., Rose, A., Frey, J.G., Brocklesby, W.S., Ladlow, M., Mellor, G.W., Bradley, M. 2001. Site distribution in resin bead as determined by confocal Raman spectroscopy. *Chem. Eur. J.* 7, 3880.

Kress, J., Zanaletti, R., Rose, A., Frey, J.G., Brocklesby, W.S., Ladlow, M., Bradley, M. 2003. Which sites react first? Functional site distribution and kinetics on solid supports investigated using confocal Raman spectroscopy and fluorescence microscopy. *J. Comb. Chem.* 5, 28.

Lee, T.K., Choi, J.H., Ryoo, S.J., Lee, Y.S. 2007. Soft shell resins for solid-phase peptide synthesis. *J. Peptide Sci.* 13, 255.

Lee, T.K., Ryoo, S.J, Byun, J.W., Lee, S.M., Lee, Y.S. 2005. Preparation of core-shell-type aminomethyl polystyrene resin and characterization of its functional group distribution. *J. Comb. Chem.* 7, 170.

Letsinger, R.L., Kornet, M.J., Mahadevan, V., Jerina, D.M. 1964. Organic and biological chemistry: Reaction on polymer supports. *J. Am. Chem. Soc.* 86, 5163.

Leznoff, C.C., Wong, J.Y. 1972. The use of polymer supports in organic synthesis: The synthesis of monotrityl ethers of symmetrical diols. *Can. J. Chem.* 50, 2892.

Li, R., Xiao, X., Czarnik, A.W. 1998. Merrifield MicroTube™ reactors for solid phase synthesis. *Tetra. Lett.* 39, 8581.

Li, W., Yan, B. 1998. Effects of polymer supports on the kinetics of solid-phase organic reactions: A comparison of polystyrene- and TentaGel-based resins. *J. Org. Chem.* 63, 4092.

Lindsley, C.W., Hodhes, J.C., Filzen, G.F., Watson, B.M., Geyer, A.G. 2000. Rasta silanes: New silyl resins with novel macromolecular architecture via living free radical polymerization. *J. Comb. Chem.* 2, 550.

MacBeath G., Koehler A.N., Schreiber, S.L. 1999. Printing small molecules as microarrays and detecting protein-ligand interactions en masse. *J. Am. Chem. Soc.* 121, 7967–7968.

MacDonald, A.A., DeWitt, S.H., Ghosh, S., Hogan, E.M., Kieras, L., Czarnik, A.W., Ramage, R. 1995. The impact of polystyrene resins in solid-phase organic synthesis. *Mol. Diversity* 1, 183.

Malavolta, L., Oliveira, E., Cilli, E.M., Nakaie, C.R. 2002. Solvation of polymers as model for solvent effect investigation: Proposition of a novel polarity scale. *Tetrahedron* 58, 4383.

Marchetto, R., Cilli, E.M., Jubilut, G.N., Schreier, S., Nakaie, C.R. 2005. Determination of site-site distance and site concentration within polymer beads: A combined swelling-electron paramagnetic resonance study. *J. Org. Chem.* 70, 4561–4568.

Martin, G.E., Shambhu, M.B., Shakhshir, S.R., Digenis, G.A. 1978. Polymer-bound carbonic-carboxylic anhydride functions: Preparation, site-site interactions, and synthetic applications. *J. Org. Chem.* 43, 4571.

Mazur, S., Jayalekshmy, P. 1979. Chemistry of polymer-bound *o*-benzyne. Frequency of encounter between substituents on cross-linked polystyrenes. *J. Am. Chem. Soc.* 101, 677.

Mazur, S., Jayalekshmy, P. 1976. Pseudodilution, the solid-phase immobilization of benzyne. *J. Am. Chem. Soc.* 98, 6710.

McAlpine, S.R., Schreiber, S.L. 1999. Visualizing functional group distribution in solid-support beads by using optical analysis. *Chem. Eur. J.* 5, 3528.

McAlpine, S.R., Lindsley, C.W., Hodges, J.C., Leonard, D.M., Filzen, G.F. 2001. Determination of functional group distribution within Rasta resins utilizing optical analysis. *J. Comb. Chem.* 3, 1.

Meldal, M. 1997. Properties of solid supports. *Methods in Enzymol.* 289, 83.

Merrifield, B. 1984. The role of the support in solid phase peptide synthesis. *Brit. Polym. J.* 16, 173.

Merrifield, R.B. 1969. Solid-phase peptide synthesis. *Advan. Enzymol.* 32, 221.

Mitchell, A.R., Kent, S.B.H., Engelhard, M. Merrifield, R.B. 1978. A new synthetic route to *tert*-butyloxycarbonylaminoacyl-4-(oxymethyl)phenylacetamidomethyl-resin, an improved support for solid-phase peptide synthesis. *J. Org. Chem.* 43, 2845.

Mogemark, M., Gårdmo, F., Tengel, T., Kihlberg, J., Elofsson, M. 2004. Gel-phase ^{19}F NMR spectral quality for resins commonly used in solid-phase organic synthesis; a study of peptide solid-phase glycosylations. *Org. Biomol. Chem.* 2, 1770.

N'Guyen, T.D., Boileau, S. 1979. Phase transfer catalysis in the chemical modification of polymers. Part II. *Tetra. Lett.* 28, 2651.

Ogino, K., Sato, H. 1995. NMR analysis of interaction between styrene-divinylbenzene gel beads and small molecules. *J. Polymer Sci. B.* 33, 189.

Overberger, C.G., Sannes, K.N. 1974. Polymer as reagents in organic synthesis. *Angew. Chem. Int. Ed. Engl.* 13, 99.

Pickup, S., Blum, F.D., Ford, W.T., Periyasamy, M. 1986. Transport of small molecules in swollen polymer beads. *J. Am. Chem. Soc.* 108, 3987.

Pittman, Jr. C.U., Evans, G.O. 1973. Polymer-bound catalysts and reagents. *Chem. Tech.* 560.

Rademann, J., Barth, M., Brock, R., Egelhaaf, H-J, Jung, G. 2001. Spatially resolved single bead analysis: Homogeneity, diffusion, and adsorption in cross-linked polystyrene. *Chem. Eur. J.* 7, 3884.

Rana, S., White, P., Bradley, M. 2001. Influence of resin cross-linking on solid-phase chemistry. *J. Comb. Chem.* 3, 9.

Rapp, W. 1996. PEG grafted polystyrene tentacle polymers: Physico-chemical properties and application in chemical synthesis. In *Combinatorial Peptide and Nonpeptide Libraries: A Handbook.* G. Jung (ed.), 425–464. VCH, Weinheim.

Rebek, Jr. J., Trend, J.E. 1979. On the rate of site-site interactions in functionalized polystyrenes. *J. Am. Chem. Soc.* 101, 737.

Regen, S.L. 1975. Influence of solvent on the motion of molecules "immobilized" on polystyrene matrices and on glass surfaces. *J. Am. Chem. Soc.* 97, 3108.

Regen, S.L.J. 1974. Influence of solvent on the mobility of molecules covalently bound to polystyrene matrices. *J. Am. Chem. Soc.* 96, 5275.

Regen, S.L, Dulak, L. 1977. Solid phase cosolvents. *J. Am. Chem. Soc.* 99, 623.

Regen, S.L, Lee, D.P. 1977. Influence of solvent on site isolation in polystyrene resins. *Macromolecules.* 10, 1418.

Reger, T.S., Janda, K.D. 2000. Polymer-supported (salen) Mn catalysts for asymmetric epoxidation: A comparison between soluble and insoluble matrices. *J. Am. Chem. Soc.* 122, 6929.

Sanetra, R., Kolarz, B.N., Wlochowicz, A. 1985. Determination of the glass transition temperature of poly(styrene-co-divinylbenzene) by inverse gas chromatography. *Polymer* 26, 1181.

Schneider, B., Doskocilova, D., Dybal, J. 1985. Motional restrictions and chain conformation in various swollen cross-linked polystyrene gels from ^{1}H NMR line-shape analysis. *Polymer* 26, 253.

Scott, L.T., Rebek, J., Ovsyanko, L., Sims, C.L. 1977. Organic chemistry on the solid phase: Site–site interactions on functionalized polystyrene. *J. Am. Chem. Soc.* 99, 625.

Scott, P.J.H., Steel, P.G. 2006. Diversity linker units for solid-phase organic synthesis. *Eur. J. Org. Chem.* 10, 2251.

Scott, R.H., Balasubramanian, S. 1997. Properties of fluorophores on solid phase resins, implications for screening, encoding and reaction monitoring. *Bioorg. Med. Chem. Lett.* 7, 1567.

Shea, K.J., Sasaki, D.Y., Stoddard, G.J. 1989a. Fluorescence probes for evaluating chain salvation in network polymers: An analysis of the solvatochromic shift of the dansyl probe in macroporous styrene-divinylbenzene and styrene-diisopropenylbenzene copolymers. *Macromolecules.* 22, 1722.

Shea, K.J., Stoddard, G.J., Sasaki, D.Y. 1989b. Fluorescence probes for evaluation of ionic reagents through network polymers: Chemical quenching of the fluorescence emission of the dansyl probe in macroporous styrene-divinylbenzene and styrene-diisopropenylbenzene copolymers. *Macromolecules* 22, 4303.

Shi, S.S., Wang, F., Yan, B. 2007. Site-site isolation and site-site interaction: Two sides of the same coin. *Int. J. Peptide Res. Thera.* 13, 213.

Sun, Q., Yan, B. 1998. Single bead IR monitoring of a novel benzimidazoles synthesis. *Bioorg. Med. Chem. Lett.* 8, 361.

Toy, P.H., Janda, K.D. 1999. New supports for solid-phase organic synthesis: Development of polystyrene resins containing tetrahydrofuran derived cross-linkers. *Tetra. Lett.* 40, 6329.

Tripp, J.A., Svec, F., Fréchet, J.M.J. 2001. Grafted macroporous polymer monolithic disks: A new format of scavengers for solution-phase combinatorial chemistry. *J. Comb. Chem.* 3, 216.

Ulijn, R.V., Brazendale, I., Margetts, G., Flitsch, S.L., McConnell, G., Girkin, J., Halling, P.J. 2003. Two-photon microscopy to spatially resolve and quantify fluorophores in single-bead chemistry. *J. Comb. Chem.* 5, 215.

Vaino, A.R., Goodin, D.B., Janda, K.D. 2000. Investigating resins for solid phase organic synthesis: The relationship between swelling and microenvironment as probed by EPR and fluorescence spectroscopy. *J. Comb. Chem.* 2, 330.

Vojkovsky, T. 1995. Detection of secondary amines on solid phase. *Peptide. Res.* 8, 236.

Wieczorek, P.P., Ilavsky, M., Kolarz, B.N., Dusek, K.J. 1982. Mechanical behavior and structure of single beads of homogeneous and macroporous styrene-divinylbenzene copolymers. *Appl. Polymer Sci.* 27, 277.

Wilson, M.E., Paech, K., Zhou, W-J, Kurth, M.J. 1998. Solvent and reagent accessibility within oligo(ethylene glycol) ether (PEG) cross-linked polystyrene beads. *J. Org. Chem.* 63, 5094.

Wong, J.Y., Leznoff, C.C. 1973a. The use of polymer supports in organic synthesis. II. The syntheses of monoethers of symmertrical diols. *Can. J. Chem.* 51, 2452.

Wong, J.Y., Leznoff, C.C. 1973b. The use of polymer supports in organic synthesis. II. Selective chemical reactions on one aldehyde group of symmetrical dialdehydes. *Can. J. Chem.* 51, 3756.

Wulff, G., Schulze, I. 1978. Directed cooperativity and site separation of Mercapto groups in synthetic polymers. *Angew. Chem. Int. Ed. Engl.* 17, 537.

Yan, B., Fell, J.B., Kumaravel, G. 1996a. Progression of organic reactions on resin supports monitored by single bead FTIR microspectroscopy. *J. Org. Chem.* 61, 7467.

Yan, B., Sun, Q., Wareing, J.R., Jewell, C.F. 1996b. Real-time monitoring of the catalytic oxidation of alcohols to aldehydes and ketones on resin support by single-bead Fourier transform infrared microspectroscopy. *J. Org. Chem.* 61, 8765.

Yan, B., Jewell, Jr. C.F, Myers, S.W. 1998. Quantitatively monitoring of solid-phase organic synthesis by combustion elemental analysis. *Tetrahedron* 54, 11755.

Yan, B., Sun, Q. 1998. Crucial factors regulating site interactions in resin supports determined by single bead IR. *J. Org. Chem.* 63, 55.

Yan, B., Martin, P.C., Lee, L. 1999. Single-bead fluorescence microspectroscopy: Detection of self-quenching in fluorescence-labeled resin beads. *J. Comb. Chem.* 1, 78.

Yan, B., Tang, Q. 2003. Comparison of kinetics of organic reactions carried out on resin beads of different diameters. *Indust. and Eng. Chem. Res.* 42, 5964.

Yu, Z., Bradley, M. 2002. Solid supports for combinatorial chemistry. *Cur. Opin. Chem. Biol.* 6, 347.

Zalipsky, S., Chang, J.L., Albericio, F., Barany, G. 1994. Preparation and applications of polyethylene glycol-polystyrene graft resin supports for solid-phase peptide synthesis. *Reactive Polymers* 22, 243.

Zhu, X.X., Zhang, J.H., Gauthier, M., Luo, J.T., Meng, F.S., Brisse, F. 2006. Large uniform-sized polymer beads for use as solid-phase supports prepared by ascension polymerization. *J. Comb. Chem.* 8, 79.

chapter three

Solid-phase reaction optimization using FTIR

3.1   Introduction

In the rehearsal phase, optimal reaction conditions need to be selected and implemented rapidly. Currently, the products and intermediates are mostly chemically (e.g., using TFA) or photochemically cleaved from the resin and then purified for analysis with conventional spectroscopic methods. The indirect information obtained from the compound derived from cleavage is used to judge the intended chemistry on the solid support. However, the requirement for cleavage of the compound from the resin for analysis renders the method inherently destructive. The "cleave and analyze" method is a time-consuming, expensive, and laborious process. Some synthetic intermediates are not stable enough for the "cleave and analyze" protocol. It is especially a waste of time and sample in characterizing intermediates in multistep synthesis. Although photochemical linkers simplify this procedure, the cleavage yield is usually low, and the types of reactions that can be carried out on such labile linkers may be limited. The cleavage of diverse compounds from resin supports relies on the chemical linker used during the synthesis. Furthermore, the isolated yield is usually different from the on-resin yield, which is important to know for multistep synthesis. It would be much more advantageous to be able to analyze the intermediates or products while still bound to the resin. These demands have consequently stimulated the development of on-support analytical methods. FTIR is a method that analyzes synthetic product exclusively on support.

Recent progresses in analytical methods associated with combinatorial chemistry have been summarized in several recent reviews (Egner and Bradley 1997; Gallop and Fitch 1998; Balkenhohl et al. 1997; Deben et al. 1998; Shapiro et al. 1997; Yan 1998; Yan and Gremlich 1999; Schneider and Anslyn 1999; Gillie et al. 2000; Hochlowski et al. 1999; Gremlich 1998, 1999; Cin et al. 2002). This chapter is specifically limited to reviewing the application of FTIR techniques with an emphasis on the single-bead FTIR technique in the monitoring of solid-phase organic synthesis (SPOS) and the impact of these methods on the rehearsal phase of combinatorial chemistry.

Applications are illustrated with examples in this overview. Experimental details can be found in original papers listed in the reference section.

3.2   Comparison of FTIR and Raman techniques

3.2.1   Techniques

The unavailability of thin-layer chromatography (TLC) in SPOS has caused severe analytical problems in the routine quality control of solid-phase reactions. Consequently, various FTIR methods have risen in importance to meet the challenge. Peptide resins have been studied using the macro FTIR method by holding milligram amounts of swollen beads between salt windows (Narita et al. 1984, 1988, 1989) and by using the FT-Raman method (Larsen et al. 1993; Hochlowski et al. 1999; Pivonka and Sparks 2000). Analysis of solid-phase organic reactions has also been performed on 5–10 mg of resin beads using various techniques, such as the KBr pellet method (Frechet and Schuerch 1971; Crowley and Rapoport 1980; Blanton and Salley 1991; Chen et al. 1994; Moon et al. 1994; Reggelin and Brenig 1996; Deben et al. 1997), diffuse reflectance infrared Fourier transform spectroscopy (DRIFTS) (Chan et al. 1997), photoacoustic FTIR spectroscopy (PAS) (Gosselin et al. 1996) and, alternatively, FTIR microspectroscopy on a single resin bead (Yan et al. 1995; Yan and Kumaravel 1996; Yan et al. 1996a, b; Pivonka et al. 1996; Pivonka and Simpson 1997).

Organic reactions selected for assembling a combinatorial library should preferably be robust and well behaved. The resin-bound structures to be monitored are thus rarely unknowns. Consequently, the major analytical task is to confirm the intended chemistry rather than to create a full structural elucidation. Because FTIR is a technique sensitive to organic functional group changes, it is well suited for this purpose. The principle of monitoring reactions by IR is based on the functional group interconversions via chemical reaction or by the appearance or disappearance of functional groups carried by building blocks or protecting groups introduced or removed during the reaction. The functional group to be monitored need not be directly involved in the reaction that is monitored. For the rehearsal of a library synthesis on solid supports, building blocks to be used in a solid-phase reaction can even be selected to contain an IR detectable group at a remote site. Many FTIR and Raman methods have been used to analyze SPOS. These techniques are described briefly.

1. FTIR microspectroscopy. This method records spectra from a small sample (such as a single-bead) in the transmission or reflectance mode. A microscope accessory is required for the measurement. A liquid-nitrogen-cooled MCT detector is used to enhance sensitivity. In the transmission mode, several beads are spread on a diamond window

and flattened by a hand-press or with a compression cell (SpectraTech, Shelton, CT). The IR beam is focused on a flattened single bead using the view mode of the microscope. The blank area surrounding the bead is cut off using an adjustable aperture. A spectrum is recorded with 32 scans (< 1 min). Then the focus is moved to the nearby blank area of the same size on the diamond window to record the background.

2. Beam condenser. This simple optical accessory is used to reduce the IR radiation from a typical 8-mm beam to 2 mm at the sample plane. This allows the analysis of the smallest quantity of pressed resin beads (50–100 beads) without using the KBr dilution. A diamond compression cell is used to flatten the beads and for the measurement. The same diamond cell without beads is then used to record a background spectrum.

3. Attenuated total reflection (macro- and micro-ATR). Under certain conditions, infrared radiation passing through a prism made of a high refractive index infrared-transmitting material (ATR crystal) will be totally reflected internally. If a sample is brought in contact with the totally reflecting surface of the ATR crystal, an evanescent wave in the less dense medium extends beyond the reflecting interface, and the evanescent wave will be attenuated in regions of the infrared spectrum where the sample absorbs energy. The intensity of the wave decays exponentially with the distance from the surface of the ATR crystal. The distance, which is on the order of micrometers, makes ATR generally insensitive to sample thickness, allowing for the analysis of thick or strongly absorbing samples. Micro-ATR measurement, which is based on the use of a micro-ATR crystal in an ATR objective, on the partial surface from a single-bead has been reported (Yan et al. 1996a). This method suffers from the lower signal intensity at the regular resin loading. One study using a very high loading resin (3.4 mmol/g) showed a better signal (Huber et al. 1999).

4. Macro- and single-bead FT Raman spectroscopy. FT Raman spectroscopy is based on inelastic light scattering, in which scattered photons exchange energy with the sample. Most commonly, the scattered photon loses energy to a vibrational mode of the sample molecule, leading to a downward frequency shift. This Raman shift is equal in energy to the light absorbed by the same molecule in an IR absorption experiment if the mode is active in both techniques. Because of the differences in the selection rules, both FTIR and FT Raman spectra must be measured to obtain the complete vibration characteristics of totally symmetric molecules or functional groups. For an asymmetric molecule, an FT Raman spectrum provides information similar to that in an FTIR spectrum, the difference being the intensity.

The factor contributing to the intensity of the Raman band is the change of polarizability during the vibration. Therefore, strong

FT Raman signals can be obtained for functional groups with low polarity and high polarizability, such as $C≡C$, $C=C$, $N=N$, S-S, C-H, S-H, $C≡N$, $C=N$, $C=S$, C-S, and symmetric vibrations of groups with a high degree of symmetry, such as NO_2, SO_2, CO_2 and aromatic rings. Because band intensity is enhanced by conjugation, FT Raman is also useful for confirming the presence of conjugated groups.

The excitation light used in FT Raman is in the near-infrared region (1064 nm). The spectra produced are free of the fluorescence interference that is usually typical for spectra obtained by visible laser excitation. FT Raman spectra can be acquired in macromode on 0.2-mg beads or in micromode such as in single-bead FT Raman microspectroscopy. Besides batch-mode bead analysis, online reaction monitoring was also demonstrated (Pivonka and Sparks 2000). An automated system has been set up with the ability to acquire ~100 spectra in an unattended overnight collection. Deconvolution for large mix and split libraries was also attained by the direct detection of library compound monomers, supported by their Raman chromophores (Hochlowski et al. 1999). Good Raman chromophores include aromatics, heteroaromatics, ethers and thioethers, and compounds containing multiple bonds between any two atoms. Stable, hour-long observations of chemical reactions on individual, trapped silica particles have been reported (Houlne et al. 2002). Using two-dimensional least-squares methods, the Raman spectra collected during the process of a reaction are resolved into their component contributions. Optical trapping combined with Confocal Raman microscopy can provide an understanding of the surface chemistry of the substrates and their effects on reaction kinetics.

5. KBr pellet method. This is a widely used sampling technique for solid samples. The method consists of mixing finely ground (0.5 μm average particle size) sample with a pure, dry spectroscopic grade of KBr powder, usually at a concentration of about 1% sample in KBr, and transferring the mixture to a die, where it is pressed until the KBr particles coalesce into a clear disk. The disk is measured using the regular FTIR instrument in which the diameter of the IR beam is 8 mm. Concerning resin samples, although polystyrene-PEG resins can be ground into small particles, polystyrene resin beads cannot. The sample particle size in the resin KBr pellet is 50–150 μm rather than the required 0.5 μm. This causes problems such as light scattering in the resin spectrum when KBr pellet is used. Recently, the use of a thin layer of KBr instead of air space in the reference cell has been shown to slightly improve the quality of spectra (Liao et al. 2000).

6. Diffuse reflectance infrared Fourier transform spectroscopy (DRIFT). When infrared radiation is directed onto the surface of a solid sample, two types of energy reflectance can occur. One is specular reflectance,

and the other is diffuse reflectance. The specular component is the radiation that reflects directly off the sample surface (it is the energy that is not absorbed by the sample). Diffuse reflectance is the radiation that penetrates into the sample and then emerges. A diffuse reflectance accessory is designed so that the diffuse reflected energy is optimized and the specular component is minimized. The optics collect the scattered radiation and direct it to the infrared detector. Typical spectra of PS and PS-PEG resins recorded with the DRIFT method as compared with the KBr method are shown in Figure 3.1.

7. Photoacoustic spectroscopy. When modulated infrared radiation is absorbed by a sample, the substance heats and cools in response to modulated infrared energy impinging on it. The heating and cooling is converted into a pressure wave that can be communicated to a surrounding gas and can be detected by an acoustic detector (essentially a sensitive microphone in the enclosed sample chamber). In such measurements, the acoustic detector replaces the infrared detector of the spectrometer. A typical spectrum recorded with this method is compared with the KBr method in Figure 3.2.

Resin beads contain 99% of the functional groups in the bead interior. A transmission measurement is a method predominantly measuring

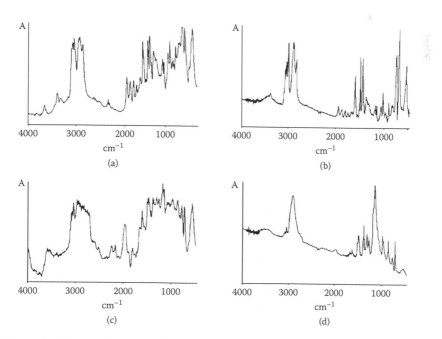

Figure 3.1 Comparison of DRIFTS and KBr spectra of NovaBiochem HL resin (a, b) and NovaBiochem TG resin (c, d). (Reproduced with permission from *Tetra. Lett.* 1997, 38, 2822–2823. Copyright© 1997, Elsevier Science Ltd.)

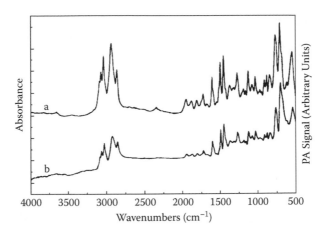

Figure 3.2 FTIR spectra of Merrifield resin (a) PA-FTIR; (b) FTIR of KBr pellet. (Reproduced with permission from *J. Org. Chem.* 1996, 61, 7980. Copyright© 1996, American Chemical Society.)

the bead interior. Of the methods described above, 1, 2, and 5 are transmission analysis methods. These methods obtain chemical information from overall substances attached to the whole bead, while 3, 6, and 7 sample only the molecules attached to the bead surface. Method 4 is based on the analysis of light scattering mostly from the outer layer of the bead.

8. Near IR spectroscopy (NIR). An NIR spectrum consists of overtones and combinations of fundamental vibrations observed in the near-infrared region; consequently, peaks are less intense, overlap, and are broader than mid-IR spectrum. An NIR reflectance method was briefly introduced as a noninvasive method in the monitoring of solid phase reactions (Hammond et al. 1999).

9. Parallel measurements. While effective synthetic schemes have been developed to generate millions of compounds, the design of efficient screening and structural characterization methodologies remains demanding and presents a bottleneck in the overall process. A large number of resin-bound compounds can be analyzed concurrently with FTIR mapping techniques (Haap et al. 1998). The automatic spectral mapping of surfaces and objects embedded in them can be achieved by the combination of an FTIR microscope with a motor-driven x-y stage. Through the mapping of larger areas, hundreds of embedded resin beads can be detected. A model library of isoxazolidines on Rink amide ARAM resin was synthesized to exhibit the potential of this method, and the typical polystyrene combination vibration at 1942 cm^{-1} was chosen for the IR reconstruction. By the superpositioning of various IR maps, compounds immobilized on single resin beads can

be characterized. Alternatively, FTIR imaging was also developed as a powerful spectroscopic tool for the parallel analysis of members of resin-supported combinatorial libraries (Snively et al. 2000). A map of about 300 different resin beads was collected, and the spectra were used to determine the character of the beads. FTIR imaging is nondestructive and highly chemically sensitive, and it provides a solution for the rapid characterization of resin-supported library members without the necessity of using invasive analytical techniques.

3.2.2   Analysis of resin-bound compounds

In order to evaluate these methods in the analysis of resin-based organic molecules, the resin-bound molecules in Diagram 3.1 were analyzed by various methods (Yan et al. 1999).

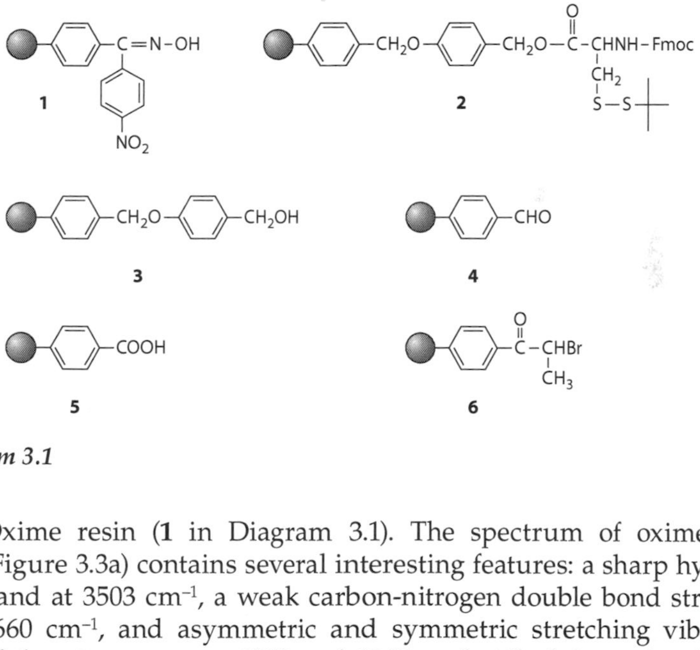

Diagram 3.1

1. Oxime resin (**1** in Diagram 3.1). The spectrum of oxime resin (Figure 3.3a) contains several interesting features: a sharp hydroxyl band at 3503 cm^{-1}, a weak carbon-nitrogen double bond stretch at 1660 cm^{-1}, and asymmetric and symmetric stretching vibrations of the nitro group at 1523 and 1345 cm^{-1}. All of these signals have been identified by FTIR methods, except for the water interference at about 3400 cm^{-1} and the more overlapping fingerprint region bands in the ATR method, as well as the usual interference and baseline tilting in the ATR method. The KBr method is subject to interference and baseline tilting. FT Raman spectroscopy failed to unambiguously detect bands at 3503 and 1523 cm^{-1}. However, the N=O symmetric stretching of the nitro group at 1345 cm^{-1} is stronger in the FT Raman spectrum than in the IR spectrum.

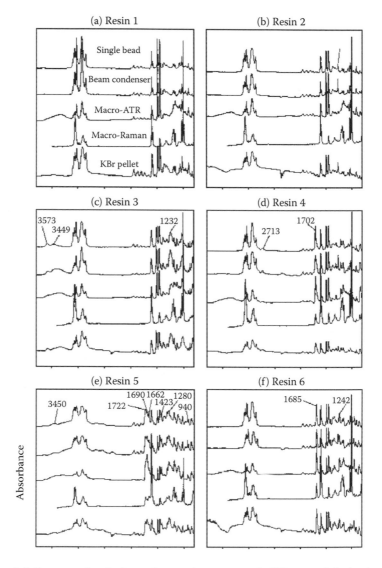

Figure 3.3 Spectra of resin-bound organic compounds (Diagram 3.1) obtained by various FTIR and FT Raman methods (see panel a). Resins are (a) oxime resin; (b) Fmoc-cys(t-buthio)-Wang resin; (c) Wang resin; (d) formylpolystyrene resin; (e) carboxylpolystyrene resin; (f) brominated Wang resin.

2. Fmoc-Cys(t-Buthio)-Wang resin (**2** in Diagram 3.1). Figure 3.3b shows the spectra of Fmoc-Cys(t-Buthio)-Wang resin obtained by various methods. Characteristic features in this compound are an ester carbonyl, an Fmoc carbonyl, an amide NH, and a disulfide bond. The two carbonyl bands overlap at 1727 cm^{-1}, and the NH

stretch is at 3420 and 3350 cm^{-1} as a doublet. These features are easily observed by FTIR methods, but not by FT Raman. The vibration of S-S is between 450 and 550 cm^{-1}, which is not detected by FTIR methods. The vibration of the S-S bond was detected by the FT Raman method (Figure 3.3b inset).

3. Wang resin (**3** in Diagram 3.1). Wang resin (Figure 3.3c) contains a benzylic hydroxyl group. A sharp band at 3573 cm^{-1} and a broad band at 3449 cm^{-1} observed by all methods with the exception of the FT Raman method are attributed to the free and intraresin hydrogen-bonded hydroxyl groups (Yan and Sun 1998). A strong CH$_2$-OH bending vibration at 1232 cm^{-1} is also observed by all methods except FT Raman, because of the different selection rule in Raman scattering. Vibrations of polar groups are not easily observable.

4. Formylpolystyrene resin (**4** in Diagram 3.1). The spectrum of formylpolystyrene resin is characterized by a band at 2713 cm^{-1} attributed to the aldehyde (O=)C-H stretch and a band at 1702 cm^{-1} attributed to the carbonyl group (Figure 3.3d). All FTIR methods can detect these signals, except for interference and S/N ratio problems in cases of ATR and KBr pellets. In the FT Raman spectrum, these signals are relatively weaker.

5. Carboxylpolystyrene resin (**5** in Diagram 3.1). Spectra of carboxylpolystyrene resin are shown in Figure 3.3e. In the liquid or in solution at concentrations over 0.01 M, carboxylic acids exist as dimers from the strong hydrogen bonding. Because of the strong bonding, a free hydroxyl stretching vibration near 3500 cm^{-1} is observed only in very dilute solution in nonpolar solvents in which a mixture of monomer and dimer is observed. In the spectrum of resin-bound carboxylic acid, the O-H shows a broad band from 3300 to 2500 cm^{-1}, a feature of hydrogen-bonded carboxyl groups. A small sharp band at 3450 cm^{-1} is also observed in spectra recorded by single-bead, beam condenser, and KBr pellet methods. This band is an indication of the presence of a portion of free carboxyl groups. This "site isolation" effect is further corroborated by the frequencies of the carbonyl group. The band at 1722 cm^{-1} was attributed to the free carboxyl groups, and the band at 1690 cm^{-1} to the associated carboxyl groups (Letsinger et al. 1964). The origin of the band at 1662 cm^{-1} is not so clear. Because this band remains after esterification of the resin, it may come from the intermediate in the resin preparation process, which is a ketoester, or an amide formed with an amine impurity during the resin preparation process. In addition to O-H and the carbonyl groups, the -OH in-plane bending at 1423 cm^{-1}, the C-O stretching at 1280 cm^{-1}, and O-H out-of-plane bending at 940 cm^{-1} are all observed. The carbonyl vibrations were weakly observed by

the FT Raman method. In addition, the above-mentioned vibrations are not observed, except for the band at 1280 cm^{-1}.

6. Brominated Wang resin (**6** in Diagram 3.1). Brominated Wang resin (Figure 3.3f) showed a characteristic peak at 1685 cm^{-1} for the ketone carbonyl group and a strong band at 1242 cm^{-1} for the Φ-C(=O) stretching and bending vibrations (Figure 3.3f), besides other features. Spectra from the single-bead and beam condenser methods clearly display these signals. ATR and KBr methods also show these features, except for the interference to which these methods are usually subjected. The macro FT Raman method showed a weak band for the benzene-conjugated carbonyl at 1685 cm^{-1} and a C-Br vibration below 600 cm^{-1} (not shown), but failed to clearly detect the Φ-C(=O) stretching and bending vibration at 1242 cm^{-1}.

3.2.3   Comparison of techniques

Spectra obtained with FTIR and FT Raman techniques presented in Figure 3.3 are compared along with the published data on DRIFT (Chan et al. 1997) and photoacoustic technique (Gosselin et al. 1996) in terms of information content and quality such as signal/noise ratio, resolution and artifact, sample requirement, speed, ease of operation, and instrument cost (Table 3.1).

1. Information content and quality. In general, all FTIR methods provide useful information on the solid-phase samples, but the quality and the extent of the revealed information vary. From Figure 3.3, it is clear that single-bead FTIR and the beam condenser methods are

Table 3.1 Comparison of Various FTIR and Raman Techniques

	Rapid Analysis	Less Sample	High S/N and Resolution	Less Artifact and Interference	Low Cost	Total
Singlebead	5	5	5	5	2	22
Beam condenser	5	3	5	5	4	22
ATR	5	3	4	3	4	19
Raman	4	2	3	5	1	15
KBr	2	1	4	1	5	13
DRIFT	5	1	2	4	4	16
Photoacoustic	5	1	3	4	3	16

Note: Various methods were compared for features listed in these five entries. Their relative rankings in these aspects were scored. The highest score is 5 and the lowest is 1.

subject to less artifact interferences and, therefore, offer more information on resin-bound compounds. This demonstrates that transmission
mode always gives the highest quality. Spectra obtained by the ATR
method exhibit lower resolution in the fingerprint region compared
with those obtained by single-bead and beam condenser methods.
Moisture interference at 3100–3500 cm^{-1} is significant in ATR spectra.
The KBr method is the only destructive method of all the methods
in this investigation. It is not possible to grind the beads. The whole
beads in the pellet cause light scattering and, thus, baseline tilting.
Moisture interference is also prominent. Because of the different selection rules for Raman spectroscopy, many important vibrations are not
detectable by the FT Raman method, such as OH, NH, aliphatic C=O,
C=N, and C-O vibrations, considering only the limited examples
studied in this work. However, FT Raman can detect some IR-inactive
or weak vibrational modes, such as the stretching of S-S (Figure 3.3b).
Thus, FT Raman can be considered a complementary technique to
FTIR methods in the analysis of solid-bound organic molecules.

2. Sample requirement. KBr pellet, DRIFT, and photoacoustic methods
generally require 5–10 mg of beads. Of these methods, DRIFT and
the photoacoustic method are not destructive to the sample. The
sample can be reused for synthesis after the analysis. This is clearly
an advantage. However, under regular SPOS scale, the reaction has
to be interrupted by analysis with these methods because of the
demanded sample size. Macro FT Raman needs about 0.2 mg of resin
beads for a 5-minute analysis (laser power 350 mW). Single-bead FT
Raman can be performed using FT Raman microspectroscopy in
10–20 min and a 1000-mW laser output. The results obtained with
both FT Raman methods are comparable (Yan et al. 1999). ATR and
the beam condenser methods require only enough beads to cover the
area of an ATR crystal (Φ 0.5 mm) and the diamond window (Φ 1.5
mm) of the compression cell. Although oversampling can ensure the
coverage, a minimum of 50–100 beads is all the amount of sample
needed. The most sensitive method is the single-bead FTIR method.
It has been demonstrated in the past (Yan et al. 1996a) that the detection limit of the method is about 125 fmol of sample, which is about
1/3840 of a single-bead loading. The amount of sample required by
single-bead FTIR, single-bead FT Raman, beam condenser, and ATR
analysis is so small that the reaction can be monitored without the
need to interrupt the reaction (real-time reaction monitoring). This is
clearly the advantage similar to what TLC offers in the monitoring of
the solution synthesis.

3. Speed of analysis. No sample preparation is required for any
method, with the exception of the KBr pellet method. All FTIR
methods exhibit high speed of analysis (1–2 minutes), except that the

KBr pellet method requires the making of pellet (~10 minutes). At a moderate laser power (~350 mW), the accumulation of an FT Raman spectrum from a 0.2-mg sample requires ~5 min. An increase in the laser power can reduce the accumulation time. However, it may cause photoreactions or photodecompositions of the sample. Single-bead FT Raman analysis requires 10–20 min and at least 1000 mW of laser power. However, the measuring time highly depends on the Raman activity of the sample. Recent developments of automated spectral acquisition and parallel acquisition have increased the speed of analysis (Hochlowski et al. 1999; Snively et al. 2000, 2001).

4. Automation. The single-bead FTIR and single-bead FT Raman methods can be automated by using a computer-controlled motorized stage (SpectraTech, Shelton, CT). Automated analysis using DRIFT can be obtained commercially (Pike Technologies, 2919 Commerce Park Drive, Madison, WI 53719; Spectros Instruments, Inc., 175 North Street, Shrewsbury, MA 01545). Macro FT Raman and KBr methods can be automated by a rotating wheel assembly. Automation of ATR and photoacoustic spectroscopy is not known to our knowledge.

5. Instrumentation cost. FTIR is the most popular analytical instrument in organic chemistry laboratories. To use the above-mentioned techniques for analyzing polymer-bound organic molecules, accessories to a regular FTIR are required. In order of increasing cost, they are KBr, beam condenser, ATR, DRIFT, photoacoustic, single-bead FTIR, macro FT Raman, and single-bead FT Raman.

6. DRIFT and photoacoustic methods. The comparison of these two methods was basis on published papers, with the assumption that authors had optimized their methods. In DRIFT measurement, several milligrams of resin was used. The analysis time was similar to other FTIR methods. The spectral quality can be seen in Figure 3.1. In general, the fingerprint region is subject to signal overlapping in DRIFT methods. The cost of the accessory is, however, low (similar to ATR).

In photoacoustic measurement, less than 10 mg of resin was used. The measurement time is a few minutes (500 scans), which is fast enough for routine applications. The spectral quality is similar to that of the DRIFT and ATR spectra. Signal overlapping in the fingerprint region is also observed (Figure 3.2). The cost of this accessory is higher than that of the DRIFT, ATR, and beam condenser.

The above comparisons are summarized in Table 3.1 in terms of numeric scores from 1 to 5. Scores based on relative comparisons in each category are summed, and the final scores are compared. In general, the most appropriate technique for a laboratory is the technique that is conveniently available. However, the value/cost ratio of these methods varies. For those chemists who are searching for a method to monitor SPOS, the comparisons offered by Table 3.1 are informative. Single-bead FTIR is an impressive method. It provides the highest quality spectrum, fast analysis, and compatibility, particularly with one-bead-one-compound libraries. The beam condenser method offers the similar quality spectrum, with a few more beads (50–100 beads). With the same speed of analysis, the operation is much simpler. Furthermore, the cost of a beam condenser is much lower than an IR microscope. The ratings of other methods are also listed.

3.2.4 Monitoring of polymer-supported organic reactions

During the process of solid-phase synthesis, size exclusion and reagent diffusion remain to be elucidated because of their effects on reaction kinetics. An infrared flow cell can trap single beads for microscopic examination. Solvent/bead partition coefficients could be obtained from aminomethyl polystyrene beads using DMF, THF, and CCl_2H_2 as solvents (Pivonka and Palmer 1999). These tools facilitate analysis and understanding of reagent penetration and intraresin reagent mobility parameters.

The single-bead FTIR method has been used extensively in kinetics studies (Yan et al. 1995; Yan et al. 1996a, b; Yan and Li 1997; Sun and Yan 1998; Li and Yan 1998). To establish the kinetics capability of the beam condenser method, the kinetics of the reaction in Scheme 3.1 were studied. The reaction of succinimidyl 6-(N-(7-nitrobenz-2-oxa-1,3-diazo-4-yl) amino) hexanoate with Wang resin produced compound 7 (Figure 3.4). The disappearance of the starting material bands at 3577 and 3458 cm^{-1} with time coincides with the formation of bands at 1732, 1580, and 3352 cm^{-1}, which are all attributable to the product. The completion of the reaction can be confirmed with confidence by the disappearance of the hydroxyl bands at 3577 and 3458 cm^{-1} (Figure 3.4). The area integrations of the emerging band at 1732 cm^{-1} were plotted against time, and the time course was fitted to a pseudo-first-order rate equation with a rate constant of 3.8×10^{-4} 1/s. Both methods showed comparable quality.

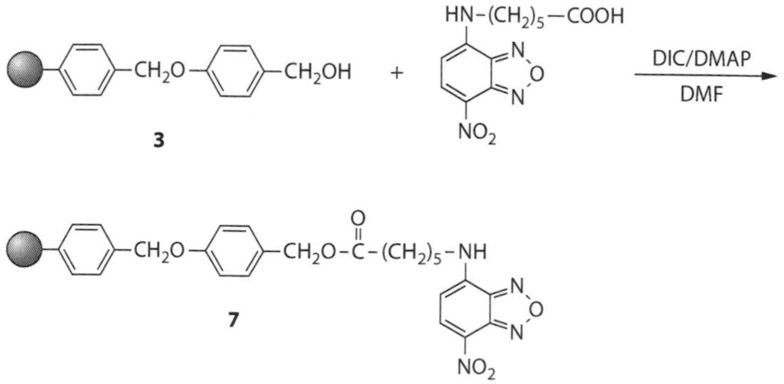

Scheme 3.1

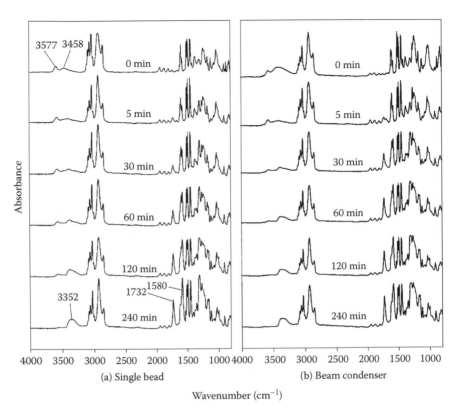

Figure 3.4 IR Spectra taken at various times during the reaction in Scheme 3.1: (a) by single-bead FTIR method; (b) by beam-condenser FTIR method.

For the purpose of confirming the desired chemistry on solid supports, FTIR is a fast, sensitive, and convenient method. In principle, all FTIR and FT Raman methods are suitable for such analysis. Although the methods can be selected through comparisons, the best method for a specific laboratory is still the method that is conveniently available and most familiar to the researcher. Because differences exist between the various methods in terms of sensitivity, speed, sample requirement, and the instrument cost, a comparison of merits of various FTIR and FT Raman methods in the context of resin bead analysis and the monitoring of solid-phase organic reactions was conducted. Both the single-bead FTIR and beam condenser FTIR methods are shown to be superb for the purpose of solid-phase reaction monitoring. The beam condenser method requires a few more beads (50–100 beads) compared to the single-bead FTIR method, but this disadvantage is well balanced by the low cost of the beam condenser and the ease of operation.

The single-bead and beam condenser FTIR methods meet several requirements as a TLC equivalent for solid-phase synthesis:

1. Only 50–100 beads are needed for analysis. The removal and examination of this tiny amount of beads from the reaction suspension practically does not perturb the reaction, so that the reaction is virtually monitored in real time, as is TLC.
2. A high-quality spectrum can be recorded within a few minutes.
3. No sample preparation is required, making the whole analysis time even shorter than a TLC analysis.
4. These two FTIR methods provide qualitative, quantitative (the percentage of conversion), and kinetics information on organic reactions carried out on solid supports.
5. From the synthetic chemist's point of view, the low cost and the ease of operation are key factors that make the beam condenser method the better method as a TLC equivalent in SPOS.

The ATR method, although not the best method for resin analysis, is the method of choice for the surface-grafted solid supports, such as multipins (Gremlich and Berets 1996) and "MicroTubes" (IRORI, San Diego, CA). More discussions on this technique will be given in section 3.5. In such analysis, it may be superior to other surface FTIR analysis methods. The FT Raman method scored low in our comparison, owing to low information content and higher cost. However, it provides complementary information to FTIR, especially for resin-bound compounds containing symmetric vibrations of organic functional groups, such as NO_2 and bands below 600 cm^{-1} such as S-S and C-Cl stretching vibrations.

3.3   The monitoring of reactions on polystyrene-based resin

A large number of SPOS reactions have been carried out on PS resins. Most on-resin analyses by FTIR techniques have also been done on PS resins. This kind of analysis will be increasingly popular and powerful in the future because the resulting quality of the IR spectra is often the same as or better than of the spectra from small molecule samples.

3.3.1   Yes-and-no information

Similar to the role of TLC in solution organic synthesis, FTIR provides a quick answer to the most important question any chemist would ask first: Is my reaction working? An example of the monitoring of a multistep indazole synthesis is shown in Scheme 3.2 (Yan and Gstach 1996). Shown in

Figure 3.5, the coupling of benzophenone-4-carboxyphenyl-hydrazone to the Merrifield resin **8** was unambiguously identified by the appearance of many spectral features in **9**. The sharp peak at 3345 cm^{-1} is assignable to the N-H stretch of the hydrazone. A very strong carbonyl band also appears at 1710 cm^{-1}, corresponding to the aromatic ester functionality introduced in this step. In addition, prominent C-O-C skeletal vibrations emerged at 1280 cm^{-1} and 1100 cm^{-1}. The quantitative oxidization of the hydrazone **9** to the (α-arylazo-benzhydryl acetate **10** by Pb(OAc)$_4$ is indicated by the complete disappearance of the N-H band due to oxidation of the hydrazone to the azo-functionality, and the appearance of a second well-resolved carbonyl stretch at 1759 cm^{-1}. The latter is assignable to the stretch of the carbonyl in the newly formed aliphatic ester. The cleavage of the N,O-acetal **10** by BF$_3$ and the successful formation of resin-bound indazole **11** is supported by the disappearance of this carbonyl band. The isolated final product was confirmed by ^{1}H and ^{13}C NMR, and the overall yield is 79%.

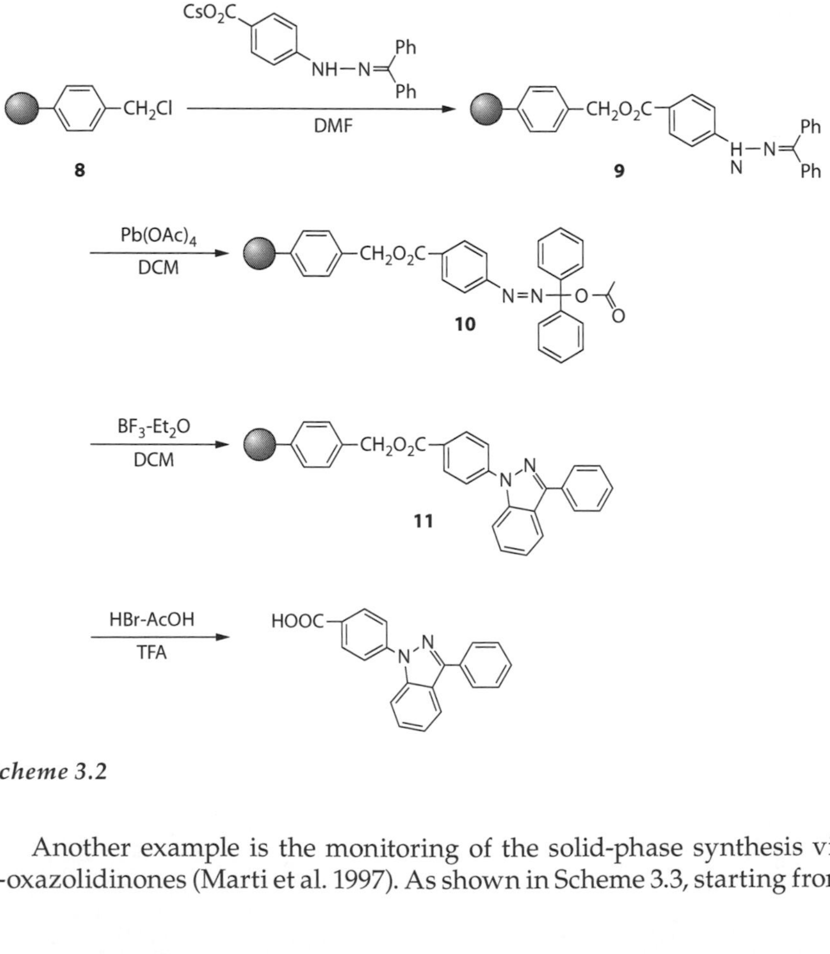

Scheme 3.2

Another example is the monitoring of the solid-phase synthesis via 5-oxazolidinones (Marti et al. 1997). As shown in Scheme 3.3, starting from

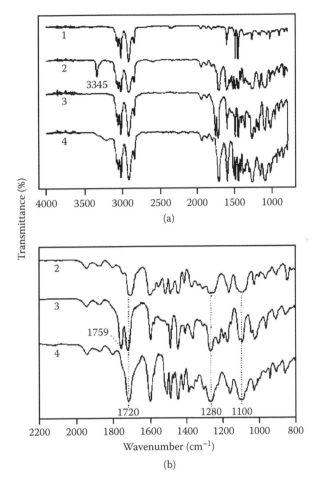

Figure 3.5 The single-bead IR spectra of starting material, intermediates, and product in Scheme 3.2. This reaction series starts with Merrifield resin. The spectra labeled 1–4 in panel a are for compounds **8–11**, respectively. Panel b shows spectra of **9–11** on an expanded scale from 800 to 2200 cm^{-1}.

the N-carbamate-protected aspartic acid, the oxazolidinone **13** was first prepared in high yield (> 90%) and purity by known procedures (para-formaldehyde, p-TsOH, azeotropic removal of water). The Alloc-protected oxazolidinones **13** were used because of the facile removal of the Alloc group by Pd^0. In the next step, the oxazolidinones **13** were attached to Knorr resin **12** using standard DIC/HOAt coupling methods. The analysis of the polymer-bound oxazolidinones **14** by single-bead FTIR microspectroscopy showed the characteristic carbonyl IR band at 1800 cm^{-1} (Figure 3.6). This unique band served as a useful and distinct indicator for product identification in a series of similar synthesis reactions.

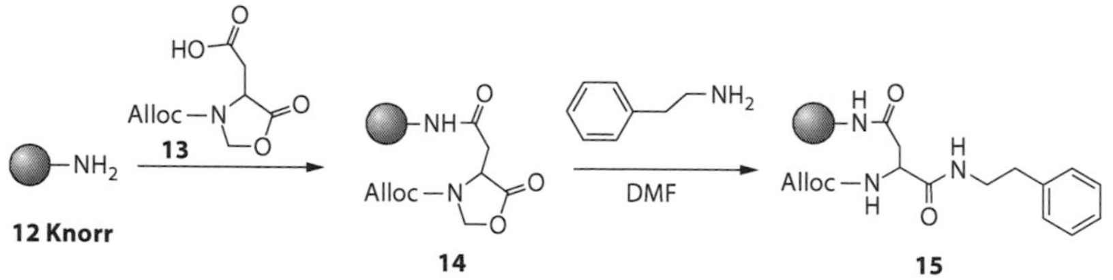

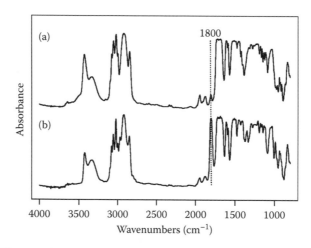

Scheme 3.3

Figure 3.6 The single-bead IR spectra for Scheme 3.3: (a) the IR spectrum of **12**; (b) the IR spectrum of **14**.

Absorbance saturation in the fingerprint region is observed for spectra in this figure. Care should be taken to make sure that the band used for analysis not be saturated. The cause for spectral saturation is the high local concentration in the bead. Bead flattening can effectively reduce the saturation effect and improve the spectrum (Yan and Kumaravel 1996). However, after the resin is treated with aqueous solution or methanol (as in Figure 3.6), the bead is shrunken and hardened and, as a result, is not easily flattened.

3.3.2   Reaction kinetics on resin support

Reactions used for library synthesis should go to completion, preferably at a high rate. This is particularly important at intermediate steps to ensure a speedy and complete reaction and a relatively high purity of the final product. To fully realize the advantage of SPOS in multistep organic synthesis, the kinetics of solid-phase reactions need to be known. However, there is a serious lack of kinetic studies for polymer-supported reactions. As a consequence, solid-phase organic reactions are commonly carried out for 12–24 hours indiscriminately before cleaving compounds off for analysis. There are two reasons for this. First, the general lack of sensitive and rapid analytical tools for on-resin analysis in the past has limited one's capability to pursue such studies. Second, most available methods (such as gel-phase NMR and most FTIR methods, such as the KBr pellet method) would require too much sample (>10 mg) for a single time point. This is impractical for general use in kinetic studies, considering the small-scale synthesis and the cost of resins. In contrast, the single-bead FTIR provides a convenient tool for kinetics studies (Yan et al. 1995; Yan et al. 1996a, b; Yan and Li 1997; Pivonka et al. 1996; Pivonka and Simpson 1997; Sun and Yan 1998; Li and Yan 1998).

To determine the reaction kinetics of the second step in Scheme 3.3, the progress of the reaction of resin **14** with a ten-fold excess of phenethylamine in DMF was monitored by taking the spectrum of a single bead at various times during the course of this reaction. Results are shown in Figure 3.7. The intensity of the band at 1800 cm⁻¹ decreases with time,

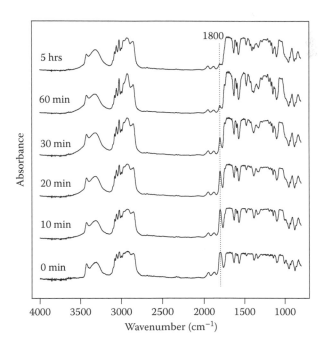

Figure 3.7 Single-bead IR spectra at various times during the synthesis of **15**. The starting resin is resin **14**.

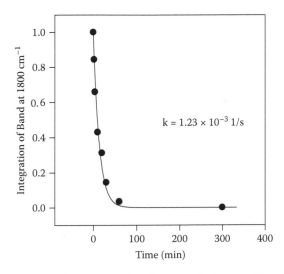

Figure 3.8 The reaction time course for the second step reaction in Scheme 3.3. The integrations of the peak area at 1800 cm^{-1} for spectra in Figure 3.5 were plotted against time.

indicating a rapid formation of the resin-bound product **15**. Integrations of areas under the oxazolidinone carbonyl band at 1800 cm^{-1} were plotted against time in Figure 3.8. Data were analyzed by curve-fitting to a first-order rate equation. The best fit is shown in Figure 3.8 as the solid line. The pseudo-first-order rate constant is 1.23×10^{-3} 1/s for the ring-opening reaction. The reaction time course is single exponential, suggesting that the rate depends only on the available reactive sites on the resin when the other reagent is in excess. Solid-phase reactions have often been fitted well to a pseudo-first-order reaction rate equation (Yan et al. 1996a). The lack of IR peaks of side products and the fit to a first-order reaction rate equation strongly suggest there are no side reactions. This was further confirmed by analyzing the cleaved product (TFA:H$_2$O, 95:5) by HPLC, MS, and NMR. The cleaved product was obtained in a "crude" yield of 80% in a purity of >90% by RP-HPLC.

In Scheme 3.4, the catalytic oxidation of resin **3** to a resin-bound aldehyde **16** was monitored by single-bead IR (Figure 3.9). The areas of IR bands that undergo changes were integrated and plotted against time (Figure 3.10). The data are fit by a pseudo-first-order rate equation to obtain rate constants as shown. The rate depends on the amount of the catalyst tetra-*n*-propylammonium perruthenate (TPAP) used for the oxidation (Figure 3.10).

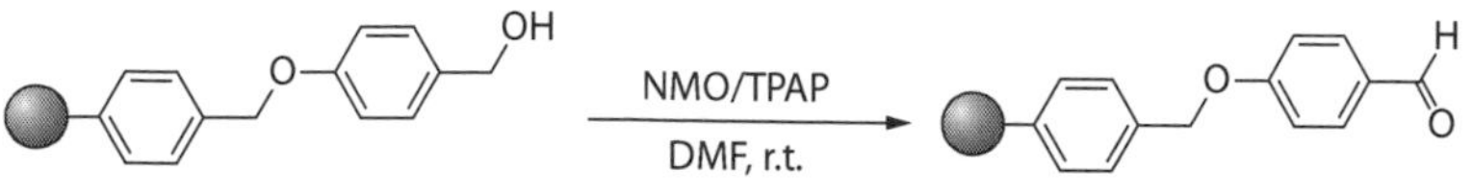

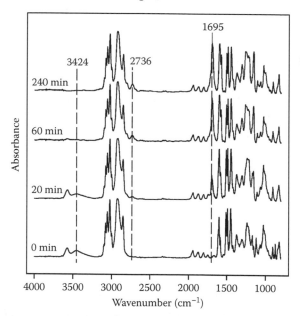

Scheme 3.4

Attenuated total reflection (ATR) or internal reflection is a versatile, nondestructive technique for obtaining IR spectra from the sample surface. Using an ATR objective in the same IR microscope for recording the single-bead transmission IR spectra, an IR spectrum from the partial surface of a single bead can be obtained within a few minutes. In order to investigate the problem of the possibly hindered diffusion of a reagent into the bead interior, the reaction rate on the bead surface and in the bead interior was monitored simultaneously using ATR mode and transmission mode, respectively (Yan et. al. 1996a). Spectra taken from a single-bead surface and interior at various times during the reaction in Scheme 3.5 are shown

Figure 3.9 IR spectra taken from a single bead at various times in the course of Scheme 3.4. Spectra were taken from a single flattened bead at 0, 20, 60, and 240 min after the initiation of the oxidation reaction. The hydrogen-bonded and unbonded hydroxyl stretch near 3400–3600 cm^{-1} disappears as the intensities of the bands for the aldehyde C-H (2736 cm^{-1}) and the aldehyde carbonyl (1695 cm^{-1}) increase.

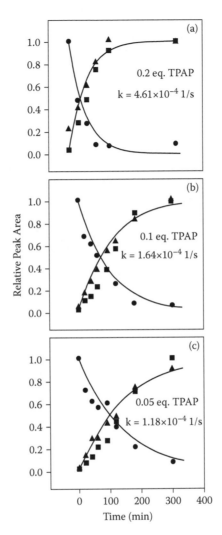

Figure 3.10 The time course of the reaction in Scheme 3.4. All spectra were normalized by making the intensity of the polystyrene band at 1945 cm⁻¹ equal. The area integration for the hydroxyl band from 3181 to 3637 cm⁻¹ (circles), aldehyde C-H band from 2664 to 2766 cm⁻¹ (squares), and aldehyde carbonyl band from 1641 to 1765 cm⁻¹ (triangles) for spectra at various times were plotted against time. Lines were calculated from the best fit to a first-order reaction equation with a rate constant shown. The amount of catalyst TPAP was 0.2 eq. in (a), 0.1 eq. in (b), and 0.05 eq. in (c).

in Figure 3.11. Reaction time courses on the bead surface and in the bead interior for two other reactions (Scheme 3.6 and 3.7) were also studied (Yan et al. 1996a), and no difference was found in the reaction rate on the surface or in the interior of the bead for these reactions.

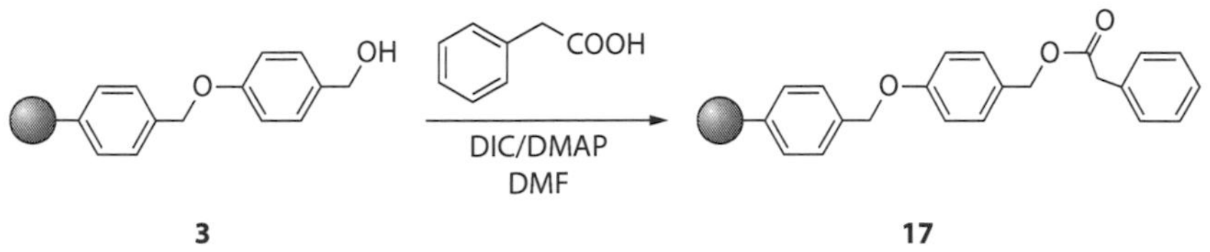

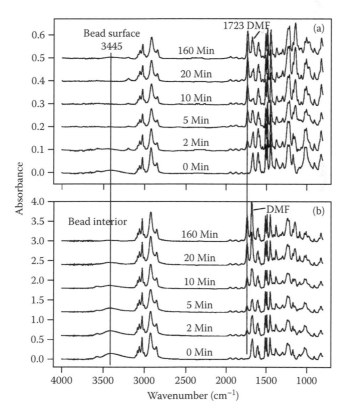

Scheme 3.5

Figure 3.11 IR spectra from a single bead taken at specified times during the course of the reaction in Scheme 3.5. Spectra taken from a partial surface of a single bead are shown in (a) and from the whole bead (99% is bead interior) in (b). The amount of compound sampled is calculated to be 130 fmol for the surface measurement and ~500 pmol for the whole bead experiments. IR absorbance bands attributable to disappearing hydroxyl group at 3445 cm⁻¹ and the emerging carbonyl group at 1738 cm⁻¹ are highlighted with the lines.

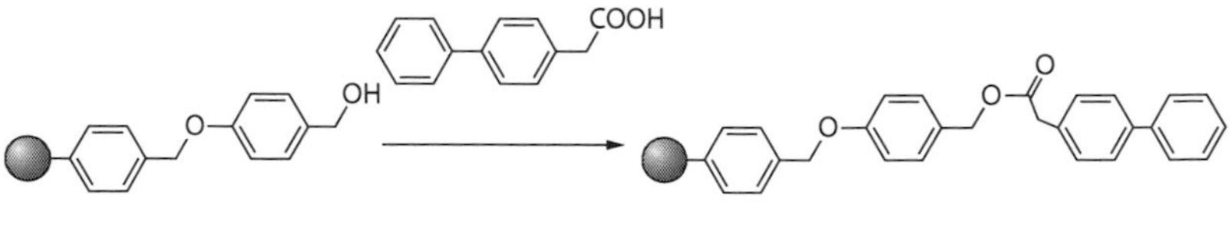

Scheme 3.6

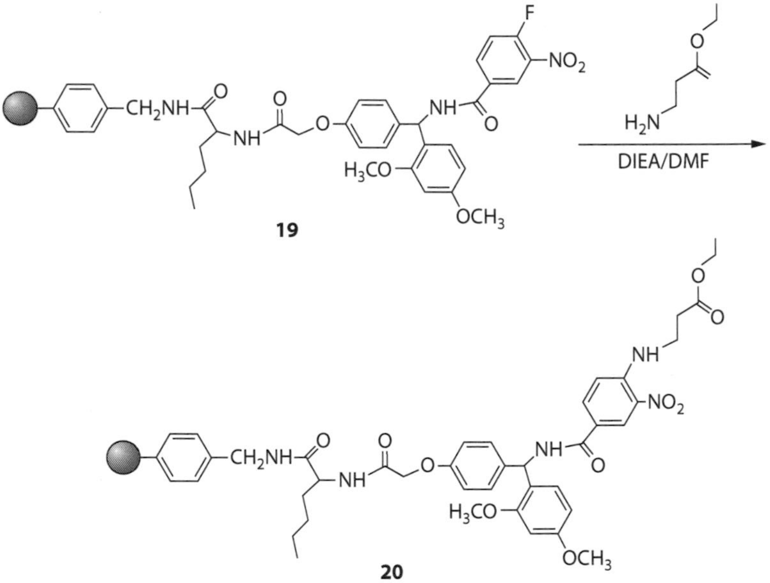

Scheme 3.7

Using the combination of FTIR microspectroscopy and a flow-through cell, organic reactions on a single flattened bead can be monitored *in situ* in the selected IR-friendly solvent (such as $CHCl_3$ or CH_2Cl_2). This instrument setup has been used to monitor a benzoylation reaction of the aminomethylstyrene resin and a series of coupling reactions to the polystyrene tritylchloride resin (Pivonka et al. 1996).

The *in situ* synthesis and analysis of compounds prepared for solid-phase synthesis were documented. Pathlength stability of the *in situ* analysis, combined with dynamic spectral processing using single-beam data collected internal to the run as spectral backgrounds, has been shown to provide very high quality spectra with enhanced spectral selectivity tailored to the problem at hand.

The feasibility of using infrared spectroscopy for the realtime, *in situ* analysis of resin-bound photocleavable linkers from a single bead is also illustrated using infrared microscopy (Pivonka and Simpson 1997).

Figure 3.12a illustrates the nitrile band absorbance region of compound **21** for a time-resolved spectral series collected from the photocleavage reaction (Scheme 3.8). Figure 3.12b illustrates the first-derivative data calculated from the Figure 3.12a spectra. First-derivative data effectively eliminate baseline drift without the need for baseline correction.

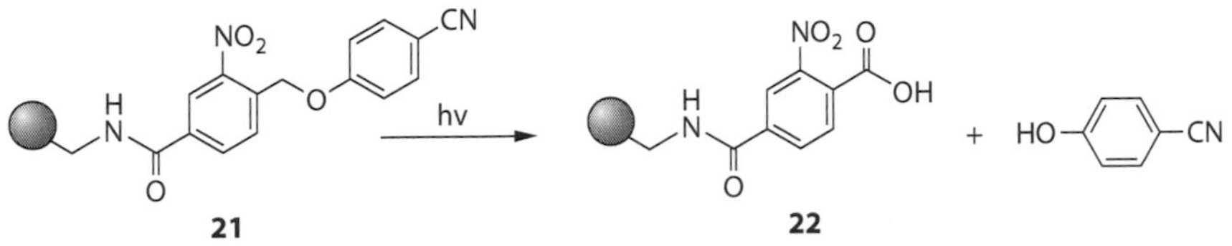

21   **22**

Scheme 3.8

This work showed that *in situ* infrared analysis of the reaction on the solid-phase support involves the single-bead analysis of photolinker cleavage kinetics. The basic methodology presented by analysis of this reaction provides a pathway for the investigation and optimization of photocleavage parameters (solvent, temperature, wavelength, and intensity). This method also allows for a rapid measurement of relative $t_{1/2}$ values for various types of groups to be cleaved in preparation for combinatorial library synthesis. In addition, the intrabead reagent dynamics has also been studied using IR spectroscopic techniques (Pivonka and Palmer 1999).

3.3.3   Quantitative estimation of chemical conversions (%) on resin

In a typical TLC experiment, if the spot of a starting material disappears and the product spot is formed, the chemical transformation can be estimated as completed. In a similar fashion, chemical conversions (%) can be conveniently estimated by single-bead FTIR analysis by observing the disappearance of the starting material band quantitatively by peak area integration (see Figures 3.5 and 3.7 for examples). All of the quantitative analyses, as are the previously discussed kinetics studies, are pathlength and bead-size independent by using a polystyrene IR band as internal reference for quantitation.

The percentage conversion of a solid-phase benzimidazole synthesis (Scheme 3.9) was monitored by the single-bead IR method (Sun and Yan 1998). The IR spectra, taken at various times during the first step in Scheme 3.9, are shown in Figure 3.13a. Among other spectral features, two distinct IR bands in compound **23** are introduced: an ester carbonyl band at 1720 cm^{-1} and an aldehyde carbonyl band at 1703 cm^{-1}. At the same time, the hydroxyl band at ~3500 is diminishing. In about 20 minutes, the intensity of the hydroxyl band almost reaches zero, and those of the carbonyl bands reach a constant maximum value, suggesting a quantitative conversion.

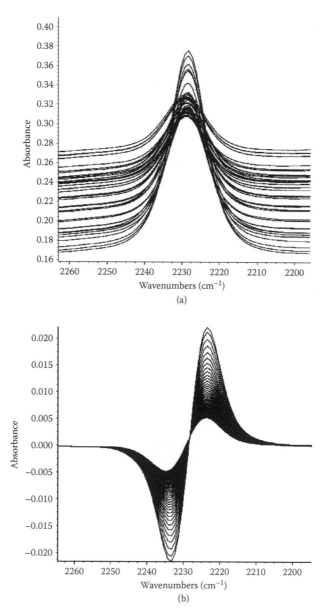

Figure 3.12 (a) Time-resolved data of the CN region of resin **21**; (b) The first derivative data from (a) (Reproduced with permission from *Anal. Chem.* 1997, 69, 3852. Copyright© 1997, American Chemical Society.)

The time course of this reaction is shown in Figure 3.13b. The yield of 97% for this transformation was independently determined directly on the resin using a 6-mg resin sample and a novel fluorescence method (Yan and Li 1997). A series of benzimidazoles, **24a–24e**, were then synthesized. The reaction was carried out at 130°C for eight hours using nitrobenzene as solvent without stirring. The success of this reaction is confirmed by the disappearance of the aldehyde carbonyl band at 1703 cm^{-1} (Figure 3.14) and the formation of the N-H stretch for **24a–d** at ~3400 cm^{-1}. Because of the overlap between the ester and aldehyde carbonyl bands (Figure 3.14), the ratio of the two bands (1703/1720 cm^{-1}) in compounds **24a–e** was determined by peak-fit analysis (for example, see Figure 3.19 later) to obtain the on-resin purity and the yield of the product.

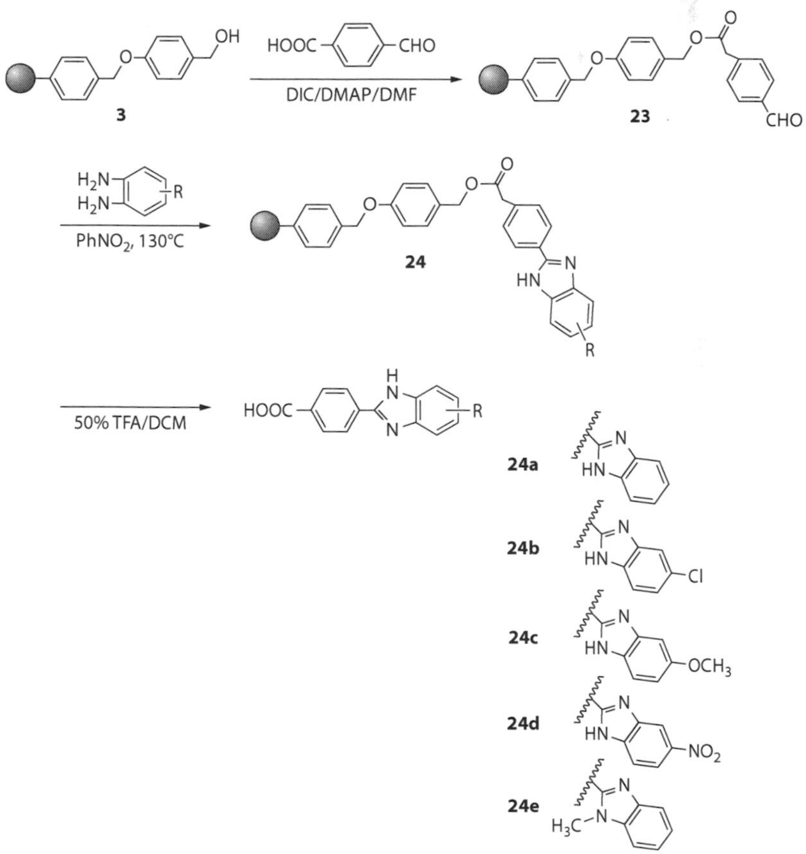

Scheme 3.9

Another example is the catalytic oxidation of a secondary alcohol to a ketone. In solution, the oxidation of a secondary alcohol to a ketone by

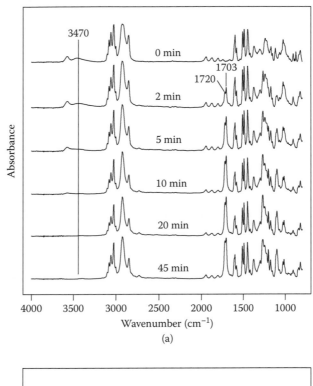

(a)

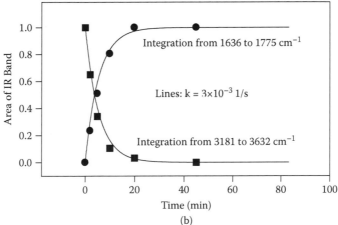

(b)

Figure 3.13 (a) Single-bead IR spectra at various times during the synthesis of **23**. The starting material is Wang resin. (b) The time course of the reaction for the synthesis of **23**.

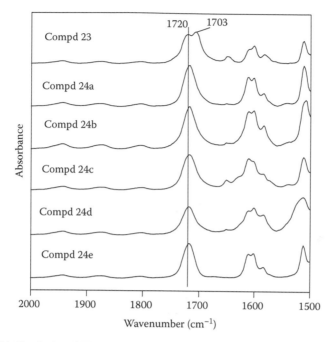

Figure 3.14 Single-bead IR spectra for resin-bound product **23** and **24a–24e**. Five sealed tubes with dried resin (100 mg, 1.0 mmol/g), one of five diamines a–e (1 mmol) and 3 ml nitrobenzene were heated at 130°C overnight. After they were cooled to room temperature, the resins were transferred to the filtration tubes and washed. Single-bead IR spectra were taken for each product.

tetra-*n*-propylammonium perruthenate (TPAP) is generally slower than the oxidation of a primary alcohol, suggesting the catalyst is sterically demanding. The sterically hindered secondary alcohol **25** in Scheme 3.10 would be expected to be further hindered by the attachment to the polystyrene resin. It is important to know the rate of this reaction. The starting material Rink acid was treated with 0.2 eq. TPAP in the presence of N-methylmorpholine N-oxide (NMO). IR spectra taken on a single bead from the reaction mixture at various time intervals after the initiation of the reaction are shown in Figure 3.15. The disappearance of a band at 3568 cm^{-1} attributed to the hydroxyl group and the increasing band at 1654 cm^{-1} attributed to the ketone carbonyl are evident. Areas under these two peaks, when plotted against time, fit with a first-order rate equation. The average rate constant is 1.9 x 10^{-4} 1/s. The same rate for the disappearance of the hydroxyl group and the appearance of the ketone carbonyl indicates that only one reaction (the oxidation of the alcohol **25** to the ketone **26**) is occurring. Based on the change in area integrations under the 3568 cm^{-1} band, a 97% conversion of **25** to **26** takes about 4 hours. More examples on the quantitative aspect of the single-bead FTIR can be found in related references (Yan et al. 1996a, b; Li and Yan 1998; Yan and Gstach 1996). Note that the estimation of the reaction yield with single-bead IR was obtained without the need to stop or interrupt the reaction or cleave the product from the support.

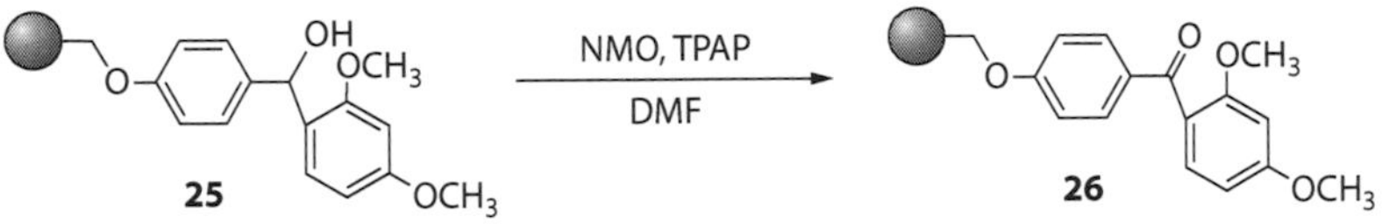

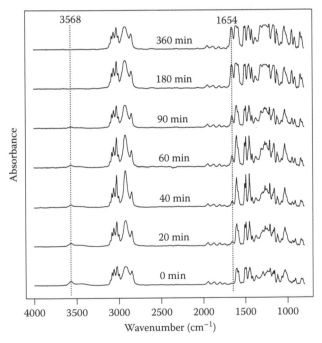

Scheme 3.10

By resorting to the isotopic synthesis and a separately predetermined standard calibration curve, reactions using isotope-labeled protecting groups or substrates have been quantitated (Russell et al. 1996). The synthesis and/or application of deuterium-isotope-containing protecting groups, including acetyl-d_3 chlorids, benzoyl-d_5 chloride, and 2-[[[(tert-butyl-d_9)oxy]carbonyl]-oximino]-2-phenylacetonitrile (BOC-ON-d_9) are also introduced as highly selective infrared signatures amenable to

Figure 3.15 IR spectra taken from a single bead at various times in the course of the reaction in Scheme 3.10. Spectra were taken from a single flattened bead at 0, 20, 40, 60, 90, 180, and 360 min after the initiation of the oxidation reaction. The hydrogen-bonded and unbonded hydroxyl stretching bands of resin **25** at 3420 and 3568 cm^{-1} disappear as the ketone carbonyl group at 1654 cm^{-1} of product **26** increase.

quantification by infrared spectroscopy. Calculation of first-derivative carbon-deuterium (C-D) stretching absorbance relative to resin backbone derivative absorbances provided pathlength-independent quantitation of deuterium content and allowed for the determination of chemical yields in solid-phase resin reactions involving changes in deuterium content. However, discrete calibrations were found to be necessary for each type of coupling: amide or ester. This technique facilitates analysis of solid-phase reactions where the number of C-D bonds changes throughout the course of a reaction. The use of deuterated protecting groups, such as Boc-d$_9$, provides for a simple method of labeling amino-containing substrates. The magnitude of the infrared absorbance from Boc-d$_9$ was found to be independent of environment; thus, the number of lysines in a multiple lysine solid-phase sample could be determined.

Calibration methods for quantifying solid-phase samples were explored by Pivonka (2000). Quantification was achieved by using solution-phase analogue spectra to calibrate solid-phase samples (so-called "analogue bleed") and using ligand-to-styrene ratios of solution-phase ligand/styrene monomers to generate internally path-length-referenced calibration. Quantification can also be achieved by solution FTIR monitoring of a byproduct generated by a solid-phase reaction (Mihaichuk et al. 2002).

3.3.4   The selection of optimal reaction conditions using single-bead IR

Because of the speed, sensitivity, and convenience of the single-bead FTIR method, the optimal solvent, resin, building block, linker, catalyst, mixing method, and other conditions have been routinely selected rapidly. The selection of a mixing method is shown below as an example.

SPOS offers several advantages over solution-phase techniques for achieving high-throughput synthesis. At the same time, SPOS also poses problems that do not exist in solution synthesis. One of the problems is that one has to select the most efficient reaction mixing method to maximize encounters between the solid-bound reactant and the soluble reagent. Diverse mixing techniques are currently used in various laboratories without a clear understanding of their relative efficiencies. Bead suspensions are usually mixed by shaking the tubes fixed at a 45° angle horizontally with an orbit shaker, by angular shaking (at 20–30°) on a wrist shaker, by rotating the tubes at an angle of 180° or 360° (whole turn) at a certain rpm, by bubbling N$_2$ gas from the bottom of the reactor, or by the conventional magnetic stirring.

To compare the mixing efficiency, the formation of a resin-bound dansylhydrazone through the reaction of formylpolystyrene resin (NovaBiochem, 0.58 mmol/g, 1% DVB, 100–200 mesh) and the dansylhydrazine was studied using six different mixing methods (Li and Yan 1997). Single-bead IR and the fluorescence quantitation method (Yan and Li 1997) were used to monitor the reaction completion. Figure 3.16 shows IR spectra from 40 and 30 individual beads after a 30-min reaction (twofold excess dansylhydrazine in DMF) while shaking with a wrist shaker or an orbit shaker. For the reaction studied, the 360° rotation and nitrogen bubbling methods provided the highest mixing efficiency under mild conditions. Wrist shaking also showed high mixing efficiency with the maximum setting (vigorous shaking). Magnetic stirring showed the same high mixing efficiency, but this method often breaks beads. The method with 180° rotation mixing required a longer reaction time and may not be preferable for slow reactions. Mixing with an orbit shaker did

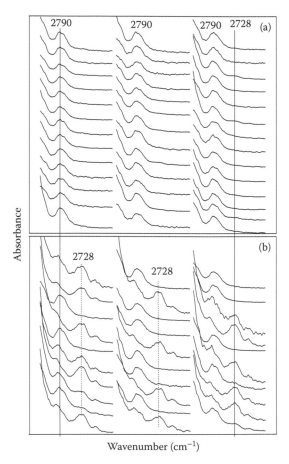

Wavenumber (cm^{-1})

Figure 3.16 IR spectra taken from individual formylpolystyrene beads after a 30-min reaction with a twofold excess of dansylhydrazine in DMF mixed with (a) a wrist shaker and (b) an orbit shaker with the reaction tube fixed at a 45° angle.

not give satisfactory reaction yields compared with the other techniques. A larger excess of reagent (ten-fold) and a doubled reaction time, however, can drive this reaction to completion.

3.4   The monitoring of reactions on PS-PEG resins: Comparing the reaction rate on PS and PS-PEG resins

3.4.1   Background

IR spectra of PS-PEG resins are easily obtained with single-bead FTIR and other FTIR techniques. The main difference between IR spectra of PS resins and PS-PEG resins is that the CH_2 vibrational band of the latter is prominent. This does not interfere with the analysis in most cases. For PS beads, the flattening of beads usually improves the spectral quality (Yan and Kumaravel 1996). PS beads can be flattened with a compression cell or simply by pressing the beads between two NaCl plates. However, the elasticity and mechanical stability of PS-PEG resin beads are so low that a slight press will shatter the bead. A new PS-PEG resin, NovaGel, has an improved elastic property and can be flattened for FTIR studies. The single-bead FTIR measurement of PS-PEG resins is usually done with the globular bead when flattening is not possible. Spectra with reasonable quality can be generated and used for the monitoring of organic reactions on PS-PEG resins.

A common problem in SPOS practice is that reaction conditions cannot simply be transferred from one kind of support to another. A set of reaction conditions may work well for polystyrene resins, but may fail completely for pin- or PS-PEG resin-based synthesis. Reaction rate may also be drastically different on different supports. Understanding the effect of various supports on reactions is critical for a successful synthesis.

PS contains mainly 1% DVB crosslinked polystyrene backbones (~96–98% by weight), which is highly hydrophobic. As a consequence of the short linker length, reactions on PS resin tend to be affected more by the hydrophobic PS backbones. The architecture of PS-PEG resin is based on a very small portion of cross-linked 1% DVB-PS backbones extensively grafted with long PEG linkers (40–60 ethylene oxide units). The PEG content is up to 70% of the resin weight. Therefore, the PEG chain determines the mechanical, physicochemical behavior of the resin. Because the reactive sites in PS-PEG resins are located at the end of long and flexible spacers and are totally separated from the PS backbone, they are not affected by PS matrix. Additionally, the dominant PEG linkers exhibit greater miscibility with most solvents, including water. PS-PEG is generally regarded as providing a quasi-homogeneous and "solution-like" reaction support. It is commonly assumed that reactions carried out on

PS-PEG resins proceed faster than those carried out on PS resins. Peptide synthesis reactions would be more favored on PS-PEG resins, because it is more polar than PS resins. However, peptide bond formation is only a special reaction. In SPOS, a different set of rules may apply.

3.4.2   Examples

To compare the reaction rate on PS and PS-PEG resins, several reactions (Schemes 3.4, 3.11, 3.12, and 3.13) were carried out on these two resins under identical conditions. An example shown in Figure 3.17 is the synthesis of a dansylhydrazone (Scheme 3.12). Resin-bound **27** reacted with dansylhydrazine under identical conditions on PS-PEG and PS resins. Single-bead IR spectra for the transformation from **27** to **28** is monitored. The ester carbonyl bands at 1734 cm^{-1} remain, while the ketone carbonyl band at 1716 cm^{-1} disappears accompanying the formation of dansylhydrazone. The formation of the band for hydrazone N-H stretch at 3230 cm^{-1} and the band for N(CH$_3$)$_2$ stretch on the dansyl group at ~2790 cm^{-1} on PS resin product proved the product formation. Although the product band at 2790 cm^{-1} is not observed because of the overlaps between the strong CH$_2$ band and the hydrazone signal in PS-PEG resin, the appearance of the band at 3230 cm^{-1} and the disappearance of the band at 1715 cm^{-1} are evident.

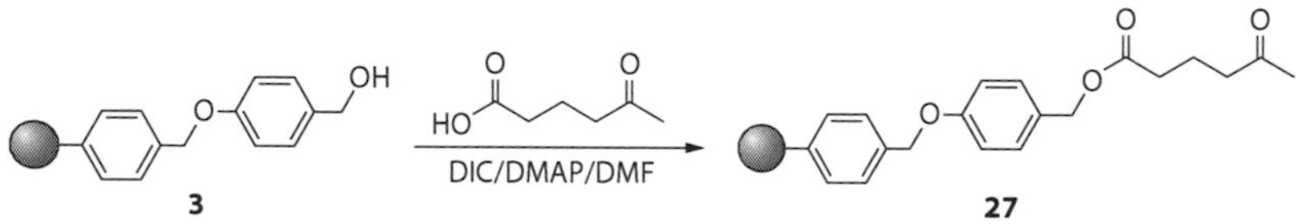

Scheme 3.11

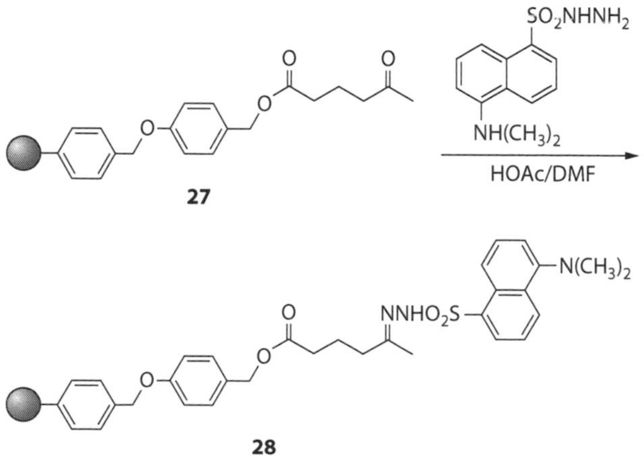

Scheme 3.12

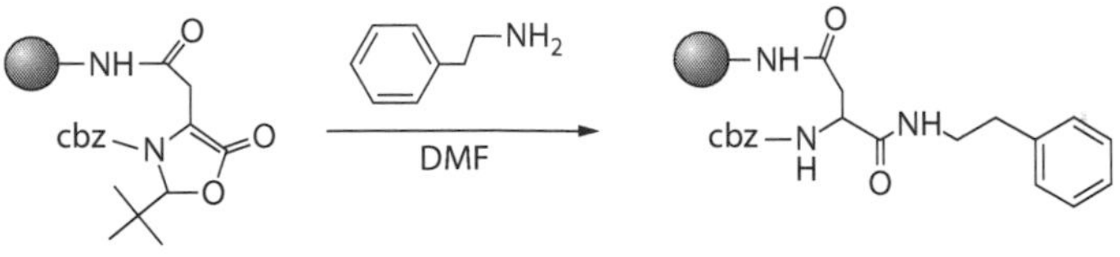

29   **30**

Scheme 3.13

Because of the overlap between the ester and the ketone carbonyl bands, a peak deconvolution analysis using PeakFit (Jandel Scientific, San Rafael, CA) was carried out for a clear-cut quantitation. As an example for the fitting method, the fitting of the data for PS reaction at 0 and 60 min (taken from Figure 3.17) is shown in Figure 3.18. It demonstrates that overlapping bands can be resolved to aid a quantitative analysis. The areas of the resolved ketone carbonyl band after PeakFit analysis was used to obtain the time course for these reactions. Results of area integrations are plotted in Figure 3.19. In this case, the reaction on PS resin is 2.2 times as fast as that on PS-PEG resin. Reaction in Scheme 3.13 was carried out on PS and PS-PEG resins. The conversion of **29** to **30** was shown in spectra taken at various times (Figure 3.20), and the time courses for reactions on PS and PS-PEG resins are shown in Figure 3.21.

The results from studies on other reactions (Scheme 3.4 and 3.11) are listed in Table 2.1 in Chapter 2. The effect of polymer matrix on the reaction rate is similar to the effect of a solvent on a solution-phase reaction rate. These data are consistent with the conclusion reached by Czarnik (1998) that solid supports are like solvents. The reaction rate on resin supports should be proportional to the compatibility of the reagent with the solvent and the resin, and to the adsorption property of the reagent into the resin. The selective adsorption of a compound into the resin depends on its similarity to the resin in terms of molecular interactions, such as hydrogen bonding, hydrophobicity, and polarity. There is no single polymer support that favors all reactions. Depending on whether the SPOS requires polar or nonpolar medium, PS-PEG- or PS-based resins, respectively, should be chosen.

3.4.3   Examples: Peptide secondary structure on resin support

In solution, peptides assume various secondary structures involving β-sheet, α-helix, or random coil. In a FTIR study on TentaGel resin using KBr pellet, peptide secondary structures were identified on resin support (Henkel and Bayer 1998). The formation of secondary structure on resin confirmed that resin is similar to a unique solvent phase. Since secondary structure formation depends on a solvent, the secondary structure on resin and in solution for the same peptide is probably different, although this was not well studied.

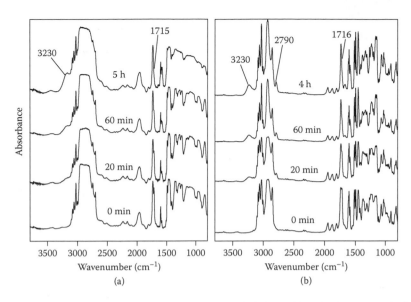

Figure 3.17 IR spectra taken from a single PS-PEG-based bead (a) or a single flattened PS-based bead (b) at various times during the reaction in Scheme 3.12.

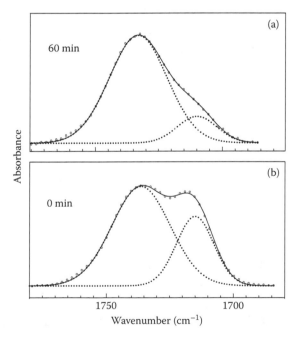

Figure 3.18 PeakFit analysis of overlapping IR bands. Overlapping IR bands at ~1716 cm⁻¹ from Figure 3.18 (b) were analyzed with the PeakFit program (Jandel Scientific, San Rafael, CA).

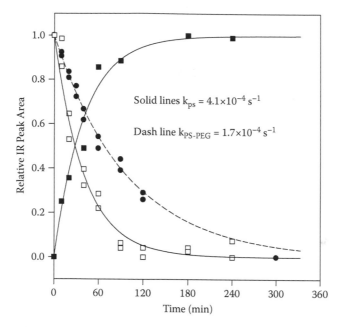

Figure 3.19 Integrations of peak areas for band at 3230 cm⁻¹ for PS-based resin (solid squares) and PeakFit resolved peak areas at 1715/1716 cm⁻¹ (open squares for PS and solid circles for PS-PEG resins). Solid lines are a theoretical time course with a rate constant of 4.1×10^{-4} 1/s, and the dashed line is a theoretical time course with a rate constant of 1.7×10^{-4} 1/s.

3.5   The monitoring of reactions on surface-functionalized polymers

Chiron "Crowns" (Maeji et al. 1995) and "Microtubes" (IRORI, San Diego, CA) are surface-coated polymers with various dimensions. Reactive groups coupled with linkers are grafted onto the polymer. In this section, the monitoring of reactions directly on crowns using macro- and micro-FTIR internal reflection spectroscopy is described.

3.5.1   Techniques

Internal reflection spectroscopy, also known as attenuated total reflectance (ATR), is a versatile, nondestructive technique for obtaining the infrared spectrum of the surface of a material or the spectrum of materials either too thick or too strongly absorbing to be analyzed by standard transmission spectroscopy.

Direct observation of the chemical synthesis on crowns is possible with FTIR accessories such as the SplitPea™ (from Harrick Scientific

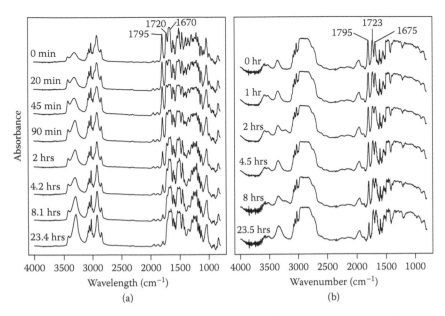

Figure 3.20 IR spectra taken from a single flattened PS bead (a) or a single PEG-PS bead (b) at various times during the reaction in Scheme 3.13.

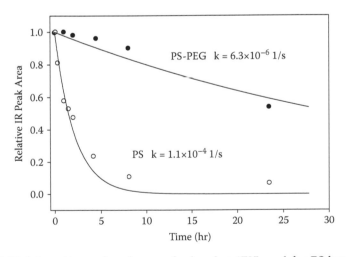

Figure 3.21 Integrations of peak areas for band at 1795 cm^{-1} for PS-based resin (open circles) and for PS-PEG-resins (filled circles). Solid lines are theoretical time courses with rate constants shown.

Corporation, Ossining, NY; see Harrick et al. 1991), the Golden Gate (Graseby Specac Inc., Fairfield, CT), and the DuraSamplIR (ASI SensiIR Tech., Norwalk, CT), which combine internal reflection with an FTIR spectrophotometer. Micro-ATR has also been used to obtain IR spectra for multipin crowns and microtubes. In internal reflection spectroscopy, no sample preparation is needed—the crown is simply clamped onto the internal reflection element. This makes it a straightforward, extremely convenient, and efficient method for the direct spectral analysis of crowns.

3.5.2 Examples

Synthesis on crowns sometimes starts with the removal of the 9-fluorenyl-methoxycarbonyl (Fmoc) protecting group from the linker. This reaction was monitored qualitatively by the macro-ATR method (Gremlich and Berets 1996).

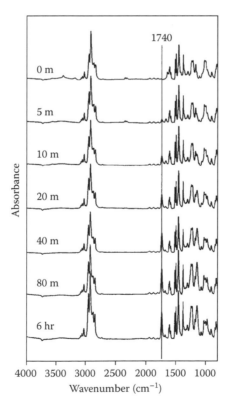

Figure 3.22 IR spectra taken from the surface of the surface of MicroTube reactors using micro-ATR FTIR at various times during the reaction in Scheme 3.5.

The reaction kinetics on the surface of the MicroTube reactor were also monitored by the micro-ATR FTIR method. Reaction in Scheme 3.5 was carried out on the surface of MicroTube reactors, and FTIR spectra were recorded at various times during the reaction (Figure 3.22). Micro-ATR studies allow a quantitative analysis of reaction kinetics on polymer surface. This reaction was found to be slightly faster on this support than on PS resins (Li and Yan 1998).

Very limited analytical methods are available for analyzing surface-modified polymer samples. The above examples show that internal reflection measurements are a convenient, quick, and reliable means for direct analysis of the surface of polymers. It is particularly advantageous for the monitoring of combinatorial chemistry reactions, which take place in a thin stratum above the "crown" or "tube" polymer. The resulting spectra, therefore, have a high information content regarding the synthesis rather than the underlying crown polymer.

With these advantages, micro- and macrointernal reflection spectroscopy are excellent tools for directly checking successive steps of a combinatorial synthesis on a polymer surface. It is extremely helpful for the elaboration of a synthesis when reaction conditions are to be optimized.

3.6   Parallel reaction monitoring

Snively and Lauterbach (2000) reported an FTIR imaging method for analyzing combinatorial libraries. They replaced the standard optics and detector used in the FTIR spectrometer with imaging optics and a focal plane array (FPA) detector. The method is a true imaging method compared to a step-and-collect mapping technique (Haap et al. 1998). This method, like other FTIR- and Raman-based methods, is a nondestructive method. The advantage of this method is that it can rapidly collect spectra from many synthetic beads and monitoring reactions in parallel (Snively et al. 2000). This method has been used in analyzing peptide libraries and catalyst libraries. Using CO adsorption on Cu-ZSM5 and Pt/SiO_2 and CO oxidation on supported Pd, Pt, Rh as model systems, it was recognized that FTIR imaging is very well suited for high-throughput parallel analysis (Snively et al. 2001). As combinatorial chemistry is used in many nonpharmaceutical areas, such as catalyst discovery, more techniques have been developed to analyze libraries and screen for active hits. IR thermography has been used to detect active catalyst in combinatorial libraries (Holzwarth et al. 1998) and to obtain time-resolved information in enantioselective catalyst libraries (Reetz et al. 1998). This method detects the IR radiation emitted by objects and has been used to monitor the activity of exothermic reactions in combinatorial libraries in parallel fashion. On the other hand, FTIR imaging can obtain chemical and structural information from the members of libraries simultaneously.

3.7 Summary

FTIR in general has been one of the most useful methods for monitoring organic reactions on solid supports. Various FTIR methods can all be used, although the amount of sample and the cost of the instrument need to be considered. As a TLC-equivalent method, single-bead FTIR is a simple, sensitive, fast, and convenient analytical method for monitoring SPOS without stopping the reaction or cleaving the product. As with TLC, single-bead FTIR provides a wide range of information, such as qualitative assessment, quantitative determination, and reaction kinetics. Studies with the single-bead FTIR have provided not only a tool for daily monitoring of the solid-phase reactions, but also a way to understand the properties of polymer-bound substrate and the nature of polymer supports and polymer-supported organic reactions. It has assisted in the rapid selection of a wide range of reaction conditions for SPOS in the rehearsal phase of combinatorial chemistry. Through its convenience and efficiency, FTIR internal reflection spectroscopy has evolved as a useful analytical methodology for monitoring combinatorial chemistry reactions directly on polymer surface.

References

Balkenhohl, C., von dem Bussche-Hunnefeld, F., Lansky, A., Zechel, C. 1997. Combinatorial chemistry. *Fresenius J. Anal. Chem.* 359, 3.

Blanton, J.R., Salley, J.M. 1991. Reducing alkyl halides using a polymer-bound crown ether/tin hydride cocatalyst. *J. Org. Chem.* 56, 490.

Chan, T.Y., Chen, R., Sofia, M.J., Smith, B.C., Glennon, D. 1997. High throughput on-bead monitoring of solid phase reactions by diffuse reflectance infrared Fourier transform spectroscopy (DRIFTS). *Tetra. Lett.* 38, 2821.

Chen, C., Randall, L. A.A., Miller, R.B., Jones, A.D., Kurth, M.J. 1994. "Analogous" organic synthesis of small-compound libraries: Validation of combinatorial chemistry in small-molecule synthesis. *J. Am. Chem. Soc.* 116, 2661.

Cin, M.D., Davalli, S., Marchioro, C., Passarini, M., Perini, O., Provera, S., Zaramella, A. 2002. Analytical methods for the monitoring of solid phase organic synthesis. *Il Farmaco* 57, 497.

Crowley, J.I., Rapoport, H. 1980. Unidirectional Dieckmann cyclizations on a solid phase and in solution. *J. Org. Chem.* 45, 3215.

Czarnik, A.W. 1998. Solid-phase synthesis supports are like solvents. *Biotech. Bioeng. (Combi. Chem.)* 61, 77.

Deben, I.E.A., Goorden, J., van Doornum, E., Ovaa, H., Kellenbach, E. 1998. Micro DRUFT: Rapid and efficient IR analysis of solid phase organic chemistry reactions. *Eur. J. Org. Chem.* 697.

Deben, I.E.A., van Wijk, T.H.M., van Doornum, E.M., Hermkens, P.H.H. 1997. Characterization of reaction products on solid supports by FT-IR. *Mikrochim. Acta. [Suppl.],* 14, 245.

Egner, B.J., Bradley, M. 1997. Analytical techniques for solid-phase organic and combinatorial synthesis. *Drug Disc. Today* 2, 102.

Fitch, W. 1998. Analytical methods for quality control of combinatorial libraries. *Annual Reports in Combinatorial. Chemistry and Molecular Diversity* 1, 59.

Frechet, J.M.J., Schuerch, C. 1971. Solid-phase synthesis of oligosaccharides. I. Preparation of the solid support. Poly[p-(propen-3-ol-1-yl)styrene]. *J. Am. Chem. Soc.* 93, 492.

Gallop, M.A., Fitch, W.L. 1997. New methods for analyzing compounds on polymeric supports. *Curr. Opin. Chem. Biol.* 1, 94.

Gillie, J.K., Hochlowski, J., Arbuckle-Keil, G.A. 2000. Infrared spectroscopy. *Anal. Chem.* 72, 71R.

Gosselin, F., Renzo, M.D., Ellis, T.H., Lubell, W.D. 1996. Photoacoustic FTIR spectroscopy, a nondestructive method for sensitive analysis of solid-phase organic chemistry. *J. Org. Chem.* 61, 7980.

Gremlich, H.-U., Berets, S.L. 1996. Use of FT-IR internal reflection spectroscopy in combinatorial chemistry. *Appl. Spectr.* 50, 532.

Gremlich, H.-U. 1998. Infrared and Raman spectroscopy in combinatorial chemistry. *Am. Lab.* Dec. 33.

Gremlich, H.-U. 1999. The use of optical spectroscopy in combinatorial chemistry. *Biotech. Bioeng. (Comb. Chem.)* 61, 179.

Haap, W.J., Walk, T.B., Jung, G. 1998. FT-IR mapping: A new tool for spatially resolved characterization of polymer-bound combinatorial compound libraries with IR microscopy. *Angew. Chem. Int. Ed.* 37, 3311.

Hammond, J., Kellam, B., Moffat, A.C., Jee, R. 1999. Non-destructive real-time reaction monitoring in solid phase synthesis by near infrared reflectance spectroscopy: Esterification of a resin-bound alcohol. *Anal. Commun.* 36, 127.

Harrick, N.J., Milosevic, M., Berets, S.L. 1991. Advances in optical spectroscopy: The ultra-small sample analyzer. *Appl. Spectr.* 45, 944.

Henkel, B., Bayer, E. 1998. Monitoring of solid phase peptide synthesis by FTIR spectroscopy. *J. Peptide Sci.* 4, 461.

Hochlowski, J., Whittern, D., Pan, J., Swenson, R. 1999. Application of Raman spectroscopy to combinatorial chemistry. *Drugs of the Future* 24, 539.

Holzwarth, A., Schmidt, H.-W., Maier, W.F. 1998. Detection of catalytic activity in combinatorial libraries of heterogeneous catalysts by IR thermography. *Angew. Chem. Int. Ed.* 37, 2644.

Houlne, M.P., Sjostrom, C.M., Uibel, R.H., Kleimeyer, J.A., Harris, J.M. 2002. Confocal Raman microscopy for monitoring chemical reactions on single optically trapped, solid-phase support particles. *Anal. Chem.* 74, 4311.

Huber, W., Bubendorf, A., Grieder, A., Obrecht, D. 1999. Monitoring solid phase synthesis by infrared spectroscopic techniques. *Anal. Chim. Acta.* 393, 213.

Larsen, B.D., Christensen, D.H., Holm, A., Zillmer, R., Nielsen, O.F. 1993. The Merrifield peptide synthesis studied by near-infrared Fourier-transform Raman spectroscopy. *J. Am. Chem. Soc.* 115, 6247.

Letsinger, R.L., Kornet, M.J., Mahadevan, V., Jerina, D.M. 1964. Organic and biological chemistry. *J. Am. Chem. Soc.* 86, 5163.

Li, W., Yan, B. 1997. A direct comparison of the mixing efficiency in solid-phase organic synthesis by single-bead IR and fluorescence spectroscopy. *Tetra. Lett.* 38, 6485.

Li, W., Yan, B. 1998. Effects of polymer supports on the kinetics of solid-phase organic reactions: A comparison of polystyrene and TentaGel-based resins. *J. Org. Chem.* 63, 4092.

Liao, J.C., Beaird, J., McCartney, M., DuPriest, M.T. 2000. An improved FTIR method for polymer resin beads analysis to support combinatorial solid phase synthesis. *Am. Lab.* Jul. 17.

Maeji, N.J., Bray, A.M., Valerio, R.M., Wang, W. 1995. Larger scale multipin peptide synthesis. *Peptide Res.* 8, 33.

Marti, R.E., Yan, B., Jarosinski, M.A. 1997. Solid-phase synthesis via 5-oxazolidino- nes: Ring-opening reactions with amines and reaction monitoring by single- bead FT-IR microspectroscopy. *J. Org. Chem.* 62, 5615.

Mihaichuk, J., Tompkins, C., Pieken, W. 2002. A method to accurately determine the extent of solid-phase reactions by monitoring an intermediate in a non- destructive manner. *Anal. Chem.* 74, 1355.

Moon, H.-S., Schore, N.E., Kurth, M.J. 1994. A polymer-supported C2-symmetric chi- ral auxiliary: Preparation of non-racemic 3,5-disubstituted-γ-butyrolactones. *Tetra. Lett.* 35, 8915.

Narita, M., Honda, S., Umeyama, H., Ogura, T. 1988. Infrared spectroscopic confor- mational analysis of polystyrene resin-bound human proinsulin C-peptide fragment: β-sheet aggregation of peptide chains during solid-phase peptide synthesis. *Bull. Chem. Soc. Jpn.* 61, 1201.

Narita, M., Isokawa, S., Honda, S., Umeyama, H., Kakei, H., Obana, S. 1989. Individuality of amino acid residues in protected peptides: Conformational and β-sheet structure-disrupted behaviors of resin-bound peptides. *Bull. Chem. Soc. Jpn.* 773.

Narita, M., Tomotake, Y., Isokawa, S., Matsuzawa, T., Miyauchi, T. 1984. Syntheses and properties of resin-bound oligopeptides: Infrared spectroscopic confor- mational analysis of cross-linked polystyrene resin bound oligoleucines in the swollen state. *Macromolecules* 17, 1903.

Pivonka, D.E., Palmer, D.L. 1999. Direct infrared spectroscopic analysis of reagent partitioning in polystyrene bead supported solid phase reaction chemistry. *J. Comb. Chem.* 1, 294.

Pivonka, D.E., Russell, K., Gero, T. 1996. Tools for combinatorial chemistry: In situ infrared analysis of solid-phase organic reactions. *Appl. Spectr.* 50, 1471.

Pivonka, D.E., Simpson, T.R. 1997. Tools for combinatorial chemistry: Real-time single-bead infrared analysis of a resin-bound photocleavage reaction. *Anal. Chem.* 69, 3851.

Pivonka, D.E., Palmer, D. 1999. Direct infrared spectroscopic analysis of reagent partitioning in polystyrene bead supported solid phase reaction chemistry. *J. Comb. Chem.* 1, 294.

Pivonka, D.E., Sparks, R.B. 2000. Implementation of Raman spectroscopy as an analytical tool throughout the synthesis of solid phase scaffolds. *Appl. Spectr.* 54, 1584.

Pivonka, D.E. 2000. On-bead quantitation of resin bound functional groups using analogue techniques with vibrational spectroscopy. *J. Comb. Chem.* 2, 33.

Reetz, M.T., Becker, M.H., Kuhling, K.M., Holzwarth, A. 1998. Time-resolved IR-thermographic detection and screening of enantioselectivity in catalytic reactions. *Angew. Chem. Int. Ed.* 37, 2647.

Reggelin, M., Brenig, V. 1996. Towards polyketide libraries: Iterative, asymmetric aldol reactions on a solid support. *Tetra. Lett.* 37, 6851.

Russell, K., Cole, D.C., McLaren, F.M., Pivonka, D.E. 1996. Analytical techniques for combinatorial chemistry: Quantitative infrared spectroscopic measure- ments of deuterium-labeled protecting groups. *J. Am. Chem. Soc.* 118, 7941.

Schneider, S.E., Anslyn, E.V. 1999. Molecular recognition and solid phase organic synthesis: Synthesis of unnatural oligomers, techniques for monitoring reactions, and the analysis of combinatorial libraries. *Adv. Supramol. Chem.* 5, 55.

Shapiro, M., Lin, M., Yan, B. 1997. On-resin analysis in combinatorial chemistry. In (ACS Book) *A Practical Guide to Combinatorial Chemistry*, Czarnik, A. W., DeWitt, S. H., eds., 123.

Snively, C.M., Oskarsdottir, G., Lauterbach, J. 2000. Chemically sensitive high throughput parallel analysis of solid phase supported library members. *J. Comb. Chem.* 2, 243.

Snively, C.M., Lauterbach, J. 2000. FTIR imaging for chemically sensitive, high throughput analysis of combinatorial libraries. In *Combinatorial Catalysis and High Throughput Catalyst Design and Testing*. Derouane, E. G., ed. 437–439, Kluwer Academic Publishers, Dordrecht.

Snively, C.M., Oskarsdottir, G., Lauterbach, J. 2001. Chemically sensitive parallel analysis of combinatorial catalyst libraries. *Cat. Today*, 67, 357.

Sun, Q., Yan, B. 1998. Single-bead IR monitoring of a novel benzimidazole synthesis. *Bioorg. Med. Chem. Lett.* 8, 361.

Yan, B. 1998. Monitoring the progress and the yield of solid phase organic reactions directly on resin supports. *Acc. Chem. Res.* 31, 621.

Yan, B., Fell, J.B., Kumaravel, G. 1996a. Progression of organic reactions on resin supports monitored by single-bead FTIR microspectroscopy. *J. Org. Chem.* 61, 7467.

Yan, B., Gremlich, H.-U. 1999. Role of Fourier transform infrared spectroscopy in the rehearsal phase of combinatorial chemistry: A thin-layer chromatography equivalent for on-support monitoring of solid-phase organic synthesis. *J. Chromatogr. B.* 725, 91–102.

Yan, B., Gremlich, H.-U., Moss, S., Coppola, G.M., Sun, Q., Liu, L. 1999. A comparison of various FTIR and FT Raman methods: Applications in the reaction optimization stage of combinatorial chemistry. *J. Combin. Chem.* 1, 46–54.

Yan, B., Gstach, H. 1996. An indazole synthesis on solid support monitored by single-bead FTIR microspectroscopy. *Tetra. Lett.* 37, 8325.

Yan, B., Kumaravel, G. 1996. Probing solid-phase reactions by monitoring the IR bands of compounds on a single "flattened" resin bead. *Tetrahedron* 52, 843.

Yan, B., Kumaravel, G., Anjaria, H., Wu, A., Petter, R., Jewell Jr., C.F., Wareing, J.R. 1995. Infrared spectrum of a single resin bead for real-time monitoring of solid-phase reactions. *J. Org. Chem.* 60, 5736.

Yan, B., Li, W. 1997. Rapid fluorescence determination of the absolute amount of aldehyde and ketone groups on resin supports. *J. Org. Chem.* 62, 9354.

Yan, B., Sun, Q. 1998. Crucial factors regulating site interactions in resin supports determined by single-bead IR. *J. Org. Chem.* 63, 55–58.

Yan, B., Sun, Q., Wareing, J.R., Jewell Jr., C.F. 1996b. Real-time monitoring of the catalytic oxidation of alcohols to aldehydes and ketones on resin support by single-bead Fourier transform infrared microspectroscopy. *J. Org. Chem.* 61, 8765.

chapter four

Reaction optimization using MS and NMR methods

4.1   Introduction

In the reaction optimization stage of combinatorial chemistry, it is essential to know whether the desired product is formed. This information is ideally obtained directly on solid support because the cleavage of the product from support is more time and labor consuming. The requirement for cleavage reaction also presents a significant limitation for any chemist who needs to either monitor a reaction in progress or characterize an intermediate in a multistep synthesis. However, the "cleave-and-analyze" method has frequently been used whenever on-support techniques are not available. This chapter will give an overview of on-support MS and NMR methods that have been used in the reaction optimization stage.

4.2   MS methods

In the solid-phase reaction optimization process, MS is a useful tool for monitoring solid-phase organic synthesis (SPOS) owing to its high sensitivity and short analysis time. Compounds from a fraction of a single-bead loading can be detected easily by electrospray ionization (MALDI) MS. A drawback of these methods is that the compound has to be cleaved before analysis. Approaches such as using photolabile linkers and using the partial cleavage by TFA vapor alleviate the problem in some cases. TFA vapor partial cleavage and MALDI-MS analysis was often used to monitor compounds attached to Wang, Rink, and HMPB linkers on a single bead or 10–100 beads (Egner et al. 1995 a, b).

MS is considerably more sensitive than any other conventional techniques, but the need for previous cleavage restricted its development. Chávez et al. (2003) developed a simple method based on mass spectrometry direct-insertion and a quadrupole detector. They set the instrument's isothermal rate at 325°C and the high-temperature and vacuum-promoted thermal cleavage on the benzylic position. Once the molecules attached to the bead were liberated, the mass spectrum could be obtained in full scan mode with the normal technique of electron impact ionization and a quadrupole detector

When synthesized on a photolabile linker (such as α-methylphenacyl-ester linker and 3-amino-3-(2-nitrophenyl) propionyl linker), a compound can be cleaved with a UV laser beam (e.g., ~337 nm) at any time during the synthesis. The released fraction of the compound can be analyzed by any conventional methods. Photocleavage combined with the MALDI-MS analysis using 1–50 beads has been used to monitor reactions (Fitzgerald et al. 1996).

Carrasco et al. (1997) has reported a construct of a photolabile linker, an ionization moiety (basic group to facilitate protonation), and a chemically cleavable linker on a support to facilitate both analysis and synthesis. Photocleavage can produce a compound that contains an ionization moiety for MS characterization on 50–100 beads, and the final chemical cleavage released the desired product.

Chu and co-workers (Yu et al. 1998) reported using MALDI-MS determination of cyclic peptidomimetic compounds on single beads. The quality control of reactions was achieved by the identification of the desired and side products. The semiquantitation of the cyclization efficiency was found to be both ring size- and sequence-dependent. MALDI-TOF-MS has been used to analyze the results of SPOS on a single bead using TFA vapor. The potential for this technique to monitor low levels of bead-surface-associated compound was demonstrated. A single component from a five-member combinatorial library was positively identified using the method (Haskins et al. 1995). The molecular weights of femtomole quantities of small peptides attached to PS beads have been determined with imaging time-of-flight secondary ion MS. The analysis is made possible by the selective clipping of the bond linking the peptide to bead with TFA vapor before the secondary ion MS assay (Brummel et al. 1994).

There have been many elegant encoding methods developed that introduce readable tags onto either the synthesis beads or the released ligands. However, those methods have many disadvantages, such as requiring extra steps to introduce tags and limiting to varying degrees the scope of the chemistry available for ligand synthesis. Using a combination of nominal mass and isotope pattern to distinguish between members of a self-coded library provides a powerful means of encoding combinatorial libraries. And such self-coded combinatorial libraries can, in principle, be applied to any screening approach where sufficient material is available for MS analysis. Although very large combinatorial libraries are required, a tagging strategy would be more appropriate (Hughes 1998).

Compared with the original methods for sequencing library-derived peptides, the partial Edman degradation method is rapid, sensitive, and inexpensive. The partial Edman degradation method converts a support-bound peptide into a series of sequence-related truncation products by repeated treatment of a support-bound peptide with a 10:1 mixture of PITC and PIC. Subsequent analysis of the peptide ladder by MALDI-MS reveals the sequence of the original peptide. But the subsequent application of this method to other peptide libraries achieved low success rates in obtaining the full-length sequences. In order to improve the reliability of the method, the PIC is exchanged with the NHS, because the NHS esters are superior to PIC as capping agents for partial Edman degradation. This provides a highly reliable, sensitive, inexpensive, and rapid method for sequencing support-bound peptides derived from combinatorial libraries (Sweeney and Pei 2003).

The development of rapid and efficient analysis for combinatorial methods to get a better understanding of the rules for functional group compatibility and substrate remains a challenge. Szewczyk et al. reports a rapid method that is a mass spectrometric labeling strategy for providing information on acceptable and unacceptable substrates, relative rates of product formation, and optimal reaction times for achieving high yields. Electrospray ionization mass spectrometry (ESI-MS) is an analysis of combinatorial libraries. After preparation of 13 authentic pyridine standards, a series of controlled experiments were performed to test the accuracy of the yields determined by MS and to test the effect of catalyst loading. Fortunately, the yields of the individual runs are in excellent agreement with those measured by ESI-MS for the same compounds when run in the library format (Szewczyk et al. 2001).

4.3   NMR methods

NMR is the primary tool for structure elucidation in low-throughput solution synthesis. However, much less information can be obtained from the NMR spectra of solid-phase samples owing to broadened spectral lines and other interference. If the purpose is only to confirm the formation of the product, NMR sometimes can be used as an analytical tool in combinatorial chemistry. The capability of NMR in structural elucidation on solid support is also under investigation in many laboratories.

In the area of drug screening, several new NMR techniques provide methods for detecting weak protein-ligand interactions that are difficult to observe using fluorescent screening approaches, yet provide an alternative starting point for drug development. These include structure-activity relationship (SAR) by NMR, NOE pumping, SHAPES, and competitive binding strategies. Advanced analysis using NMR often employs multidimensional experiments that are often time consuming. The development of new instruments, including automatic sample changers and flow probes, have improved throughput. Separated RF coils and impedance-matching circuitry are used for each sample. Chemical shift imaging is used to collect NMR spectra of up to nine samples contained in parallel capillary tubes. Recent developments in high-throughput NMR indicate that, with further development, parallel sampling NMR techniques along with high-speed multidimensional data acquisition will play an important role in high-throughput chemical analysis and drug screening (Raftery 2004).

4.3.1 Gel-phase NMR

Advancement in the development of NMR techniques for on-resin analysis has been reviewed (Egner and Bradley 1997; Keifer 1997, 1998; Gallop and Fitch 1997; Shapiro and Wareing 1998).

Gel-phase NMR has for decades been a well-established method for acquiring NMR data on solid-phase samples. Because the solvent-swollen resin slurries are neither fully solid nor fully liquid, they are referred to as gel-phase samples. The term *gel-phase NMR* now refers to the use of a conventional liquids probe (a probe in which the sample-spinning axis is aligned along the magnet bore [z] axis) to acquire the NMR data.

It is assumed that a solvent that causes more swelling will produce narrower linewidths due to increased motion within the swollen resin. Systematic studies of the influence of different solvents upon the ^{1}H NMR spectra of different resins have demonstrated, however, that this generalization is far too simplistic to account for the observed linewidths.

All substances exhibit a physical property called *magnetic susceptibility* that provides a measure of how each substance affects a surrounding magnetic field. When solution samples are analyzed, the uniform (homogeneous) substance will cause only a uniform change of the applied (static) magnetic field. This is typically corrected by "shimming" the static field of the NMR magnet to make the sample magnetically homogeneous. A resin sample is composed of discrete beads that can be described as having an organic moiety attached via a variable-length linker to a more rigid polymer matrix. Because the susceptibility of the resin and the solvent are different, solid-phase synthesis resins are heterogeneous samples. They have regions of differing magnetic susceptibility throughout the sample. This difference caused localized variations (distortions) in the magnetic field that are not correctable by "shimming" the field. The magnetic susceptibility mismatches cause additional linebroadenings that scale up with the NMR observation frequency.

Reduced molecular motion in solid samples can also broaden the NMR linewidth. ^{1}H or ^{13}C NMR spectra of the dry beads consist only of broad, unresolved resonance, as is typical of solid polymers. If the resin beads are swollen to form slurry, many lines will narrow owing to an increase in the molecular motions that allow the chemical shifts to average to their isotropic values (Rapp 1996; see Figure 4.1). While the

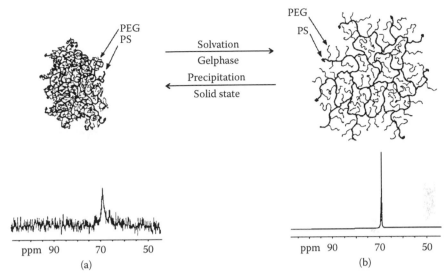

Figure 4.1 Top: resin in shrunken and swollen condition. The bottom shows the corresponding gel-phase ^{13}C NMR of the PEG resin: in swollen state, the tentacles show high T_1 values (highly flexible, well solvated) resulting in sharp ^{13}C resonance, whereas in the nonswollen state, the ^{13}C resonance shows broadening. (Reproduced with permission from Rapp, W., in *Combinatorial Peptide and Nonpeptide Libraries* edited by Jung, G., 1996, 828. Copyright© 1996, Wiley-VCH, Weinheim.)

relative rigid polymer matrix would still be expected to generate broad NMR resonances, the attached organic moieties should have sufficient molecular motion to generate narrow linewidths. Gel-phase NMR was born this way. Bayer and co-workers (1990) studied the spin-lattice relaxation times of the tripeptide Boc-Gly-Pro-Pro-OMe in solution, on soluble supports, and bound to various solid supports. The investigation revealed that the T_1 value and, hence, the mobility of the carbon atoms of the peptide esters decrease in the sequence methyl ester > PEG ester ≥ PEG-PS ester > PS ester. The results suggest that the grafted copolymer PS-PEG allows more internal molecular motion than do PS resins (Figure 4.2).

Because of linebroadening in gel-phase NMR spectroscopy, the application of gel-phase NMR is usually for lower frequency nuclei that have a wide chemical shift dispersion. For example, although the linewidth for ^{13}C spectra may be 25–75 Hz or more, the resonance is still resolved well enough to allow structural assignments. Gel-phase NMR spectroscopy has found applications for nuclei such as ^{13}C, ^{19}F, ^{2}H, ^{31}P, and ^{15}N. ^{13}C and ^{19}F NMR are most useful. Incorporation of such isotopes as ^{13}C and ^{15}N have been investigated and found very useful.

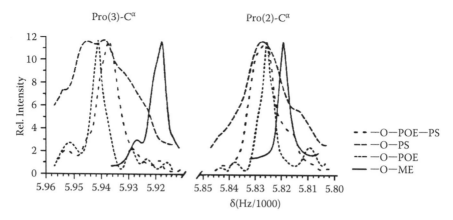

Figure 4.2 Linewidth of the signals of α-carbon atoms of Pro(2) (57.8–57.9 ppm) of the tripetide (Boc-Gly-Pro-Pro) esters (at 100.6 MHz). (Reproduced with permission from *Macromolecules* 1990, 23, 1937–1940. Copyright© 1990, American Chemical Society.)

4.3.1.1 Gel-phase ¹³C NMR

In a reaction monitoring of solid-phase peptide synthesis (Giralt et al. 1984), gel-phase sample was prepared by mixing the resin sample with $CDCl_3$ for 30 min after agitation with a vortex shaker or, sometimes, ultrasonic agitation. ¹³C NMR spectra were obtained at 50.31 MHz on a Varian XL-200 spectrometer. An array of starting resins were analyzed from a quantitative and a semi-quantitative point of view. These resins include the commonly used ones, such as Merrifield, benzhydrylamino, and Wang resins. All the carbons were assigned. A synthesis of a tripeptide, $H-Asn-(N-Me)-Ala-Thr-NH_2$, was monitored step by step. There is a general tendency for the amino acid carbons to move to a lower field when the Boc group is removed and to a higher field when the next amino acid is attached. The effect suggests the possibility of steric compression. In this monitoring procedure, the appearance of side reactions during the growing of the peptide chain can be detected.

Measurements of relaxation times were used to obtain an estimate of the mobility of the different parts of the peptide molecule at each stage, providing a quantitative basis to improve the synthetic experimental conditions. It was found that the Boc methyl groups of Boc-Thr(Bzl)-OH in solution and of Boc-tripeptide resin show similar relaxation rates. This gives a direct verification of the fact that solvent in gels in which polymer-supported peptide synthesis is carried out can be almost as efficient as solvation in solution and, therefore, accounts for the possible similar reactivity observed in solid phase and in solution.

In peptide synthesis, the coupling step can be monitored to completion by a sensitive amine test. Unfortunately, there are not many corresponding tests to check for quantitative removal of the matrix-bound peptide N-terminal protecting groups. It is inevitable that any residual protected amino groups will become exposed in a subsequent chain elongation cycle. Erroneous peptide sequences will then result. Epton et al. (1980) used gel-phase ^{13}C NMR spectroscopy in a study of polymer-supported peptide synthesis on a phenolic crosslinked poly-(acryloyl-morpholine)-based matrix. The ^{13}C NMR spectra were obtained using a Bruker NH90FT NMR spectrometer operating at 22.63 MHz, with broadband irradiation to decouple the protons. Satisfactory signal-to-noise ratio was achieved using 30000–150000 pulses at 0.7 s intervals with 0.15 g of resin swollen in 2 ml of DMSO-d_6.

Peaks due to almost all the carbons in groups pendant on the polymer backbone are readily discerned and have been assigned by comparison with the spectra of the monomers. The backbone carbons cannot be seen because (a) they fall in the 35–42 ppm region, which is obscured by DMSO-d_6; (b) the intensity will be spread over several lines as a result of tacticity effects; and (c) the line will be broadened by the relatively low mobility of the backbone.

The ^{13}C NMR spectrum of the DMSO-d_6 swollen phenolic-poly(acryloylmorpholine)-based matrix after substitution with Boc leucine (~0.5 mmol/g by amino acid analysis) and re-O-acetylation was recorded. In addition to the peaks attributable to the parent matrix, the expected leucine peaks and Boc peaks are all well resolved and readily assigned. The fact that the major sample peak in the spectrum is due to the three magnetically equivalent methyl carbons, -C(CH$_3$)$_3$, of the Boc group is most important because Boc is one of the widely used protecting groups in both SPPS and SPOS.

A step-by-step synthesis of a peptide enkephalin on polymer support was studied. As the peptide chain is lengthened, the signals from the carbons nearer the support tend to diminish in height and to broaden. The peptide unit at the end of the chain, next to Boc, always gives the sharpest lines, which is understandable in terms of decreased mobility of the nuclei nearer the support. When the next amino acid unit is added, the α-carbon signal from the unit previously adjacent to Boc tends to move upfield. This is a normal end-group effect, as mentioned previously. This example showed that the application of gel-phase ^{13}C NMR spectroscopy to SPPS yields directly hitherto inaccessible information on the peptide chain. Almost certainly, the removal of other protecting groups such as Fmoc could also be readily checked.

The application of gel-phase NMR in monitoring solid-phase organic synthesis was reported by Leznoff and co-workers (Jones et al. 1982). They reported the ^{13}C NMR spectra of a wide variety of organic substrate bound to 2% cross-linked polystyrenes. Their ^{13}C NMR spectra were determined in the Fourier mode using a Bruker HFX-270 spectrometer at 67.89 MHz. In general, spectral width of 15 kHz was observed in 16 K data points using an approximate pulse angle of 60° and averaging 2–4 K acquisitions. Delays of 10–15 s between pulses were used when integration was desired. Samples were in 10-mm tubes. Solvent-induced swelling of 0.75 g of polymers in 2–2.5 mL of CDCl$_3$ generally produced an adequate sample volume of the appropriate slurry consistency.

The results of the study, like results from SPPS synthesis, were consistent with the notion that segmental motion of side-chain substituents attached to the backbone polymer will result in sharply defined ^{13}C NMR signals for the more mobile portion of the substituent molecules. They showed that resonances of the carbon atoms close to the polymer-binding site are broadened relative to those associated with portions of the side chains considered to be more freely rotating. These indicated increasing T_1 and T_2 values for the carbon atoms of the side-chain substituents with distance from the polymer backbone. Most significant from an assignment point of view are the obvious broad lines (relative shorter T_2 values) generally associated with the first three carbon atoms of the side-chain substituents from the polymer.

Figure 4.3a shows a ^{13}C NMR spectrum of the 2% cross-linked polystyrene. The quaternary aromatic carbons (145.8 ppm), the otho and meta carbons (128.0 ppm), the para carbons (125.7 ppm), and the backbone methylene and methine carbons (40.4 and 43.0 ppm) are observed. The spectrum of polymer-bound trityl alcohol is shown in Figure 4.3b. The characteristic broad line at 81.9 ppm is attributed to the quaternary carbon atom in the triphenylcarbinol group and that at 147.0 ppm to the α-carbons in the phenyl ring.

The spectrum at the bottom of Figure 4.3 shows the ^{13}C spectrum of a polymer-bound C12 bromoalkyne polymer-Tr-O-(CH$_2$)$_2$C≡C(CH$_2$)$_8$Br derived using an alternative synthetic precursor, the monotrityl ether of ethylene glycol. The broad lines for resin backbone and carbon atoms closer to the backbone (C1-C5) and the sharper lines for carbon atoms remote from the backbone are evident in the spectrum. These results are reminiscent of the spectra of resin-bound peptides in previous discussions.

Blossey and Cannon (1990) reported that the reaction monitoring of synthetic transformations of the bound steroids containing carbonyl and hydroxyl groups and esterification of hydroxyl functionality was monitored by ^{13}C NMR spectra of solvent swollen polymer gels. The ^{13}C NMR spectra of the bound dehydrocholic acid (3,7,12-trioxo-5 β-cholanoic acid) and cholic acid (3α,7α,12α-trihydroxyl-5β-cholanoic acid) permit structural assignments to be made, on a qualitative basis, of the intact polymer without resorting to cleavage reactions of the bound materials.

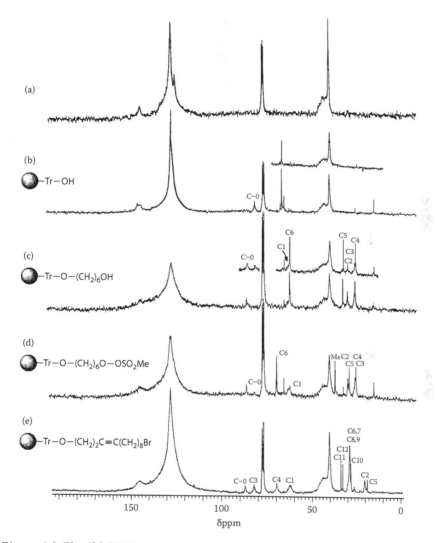

Figure 4.3 The ^{13}C NMR spectra, determined at 67.89 MHz, of solvent (CDCl$_3$) swollen slurries of (a) Eastman 2% divinylbenzene cross-linked polystyrene; (b) polystyrene-bound trityl alcohol; the inset shows the same spectrum after solvent evaporation and resuspension; (c) the polymer-bound monotrityl ether of hexanediol; the inset is the spectrum obtained using a pulse delay of 10 s; (d) the polymer-bound methylsulponate derivative; (e) the polymer-bound C$_{12}$ bromoalkyne. The side-chain carbon atoms are numbered from the trityl ether anchoring site; C-O refers to the trityl carbinol carbon, and Ms refers to the methylsulponate methyl carbon. (Reproduced with permission from Jones et al., *Org. Magn. Res.* 1982, 18, 237. Copyright© 1982, John Wiley & Sons Limited.)

There is good agreement between the literature values for solution spectrum of methyl cholate and polymer-bound cholate recorded in this study. The mono-, di-, and triesterification of the compound are determined from the ^{13}C NMR spectrum, with separation of chemical shift values for each of the carbons bearing the ester groups at 3, 7, and 12 positions. Oxidation of the polymer cholate with DMSO-oxalyl chloride (Swern oxidation) gave a quantitative yield of the triketo derivative, with a disappearance of the hydroxyl-bearing carbons at 67–70 ppm and the appearance of small signals for carbonyl carbons in the 208–212 ppm range.

By using the ^{13}C-enriched building blocks, ^{13}C NMR experiments were used to monitor SPOS on as little as 20 mg of resin (Look et al. 1994). Although labeled compounds can be quite costly, the application of this method is still possible in small-scale synthesis for reaction optimization studies. In this work, the PS-PEG resin is suspended in the NMR solvent benzene-d$_6$ and is administered to the tube insert for immediate observation. The use of ^{13}C-enriched building blocks enables one to obtain high-quality spectra at the center of interest without interference from unenriched peaks and solvent. Meaningful data can be obtained in as few as 64 transients. Typical turnaround times from preparation of the resin to the recovery of the beads for subsequent reactions are from 15 to 30 min. The enriched center need not be directly involved in the reaction that is monitored because changes in hybridization as well as electron density can be easily detected.

4.3.1.2 Gel-phase ^{19}F NMR

The first application of gel-phase ^{19}F NMR in monitoring of SPPS was reported by Manatt et al. in 1980. Proton decoupled ^{19}F 94.1 MHz NMR spectra of fluorobenzyloxycarbonyl and 2-(4-fluorophenyl)-2-propyloxycarbonyl derivatives of amino acids and peptides bonded to 1% crosslinked polystyrene were studied (starting resin "Bio Beads" S-X-1, 200-400 mesh, 1.25 mmol/g in CH$_2$Cl$_2$, 5-mm sample tubes of slurries of swollen resin, CHCl$_3$ solvent 0.5% C$_6$H$_5$F and 5% C$_6$F$_6$). Relatively narrow spectral lines were obtained (Figure 4.4). This offered the potential for studying and monitoring the SPPS. Internal concentration references are easily derived by several approaches. A ^{19}F NMR probe with a chemical shift different from that of the shift in the protecting group and inert to SPPS conditions may be coupled to a portion of the resin functional groups (Figure 4.4a). Alternately, resin beads possessing an inert ^{19}F NMR probe may be added to the resin beads on which the peptide will be synthesized (Figure 4.4b).

The time-dependent change of a nucleophilic aromatic substitution (S$_N$Ar) reaction on Rink amide resin was monitored by gel-phase ^{19}F NMR using about 1 g of resin (Scheme 4.1). Two time points, at 15 and 30 min, are shown in Figure 4.5, and the reaction was finished within 2 hours (Shapiro et al. 1996).

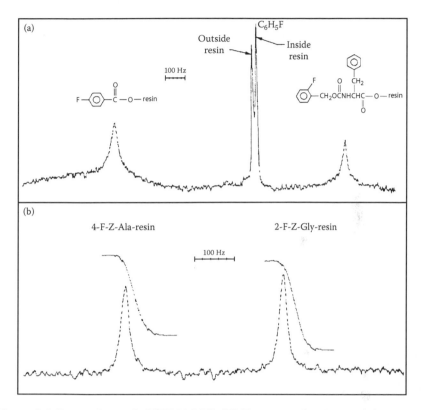

Figure 4.4 Proton decoupled ^{19}F 94.1 MHz NMR spectra of amino acid derivatives and peptides bonded to 1% cross-linked polystyrene (starting resin "Bio-Beads," s-x-1, 200–400 mesh, 1.25 mmol/g loading); 5 mm O.D. sample tubes of slurries of swelled resin; $CHCl_3$ solvent 0.5% C_6H_5F and 5% C_6F_6; for digital, 520 sec. Sweeps of 1000 Hz; for (a) sweeps of 50 s of 2000 Hz; chemical shifts in Hz from C_6H_5F: (a) 2-F-Z-Phe-resin, -477 Hz, 4-fluorobenzoic acid-resin on same resin, +704 Hz; (b) spectrum of a mixture of 27.0 mg 2-F-Z-Gly-resin beads (0.98 mmol/g) with 22.5 mg 4-F-Z-Ala-resin beads (0.91 mmol/g) with integration from 16–20 seconds sweeps. (Reproduced with permission from *Tetra. Lett.* 1996, 37, 4671–4674. Copyright© 1996, Elsevier Science Ltd.)

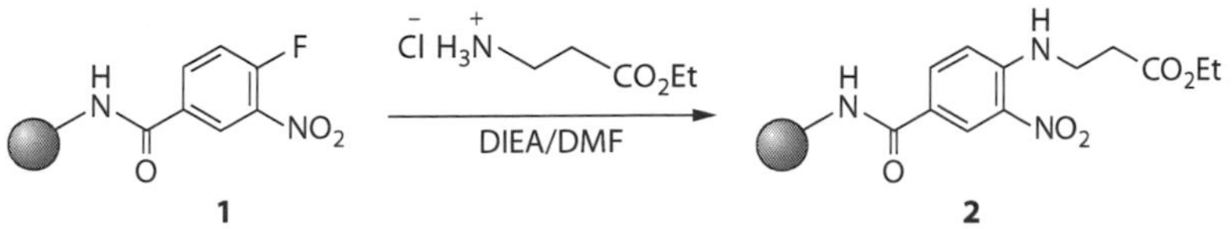

Scheme 4.1

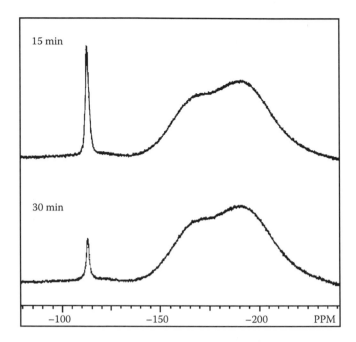

Figure 4.5 Gel-phase ¹⁹F NMR spectra of benzene-d₆-swollen resin 1 at (a) 15 min and (b) 30 min after exposure to ethyl 3-aminopropanoate-HCI/DIEA. δ(1) = –115.3 ppm. (Reproduced with permission from *Tetra. Lett.* 1996, 37, 4671–4674. Copyright© 1996, Elsevier Science Ltd.)

Gel-phase ¹⁹F NMR spectroscopy has also been used to characterize products from a variety of reactions of fluorinated aromatic compounds linked to a TentaGel resin (Svensson et al. 1996). Spectra of high quality were obtained in less than 10 min on 50–200 mg of resin. Linewidths for the compounds linked to the solid support approached those in solution and indicated that shift difference of ≤0.5 ppm can be detected. Substantial chemical shift differences were obtained for almost all of the synthetic transformations, illustrating the potential of ¹⁹F NMR for rapid monitoring of reactions in SPOS.

Ten cross-linked polystyrene-supported, protected chiral amines featuring both a spacer, comprising 5 to 15 atoms, and a fluorinated linker have been successfully prepared by Hourdin et al. (2005). The development of the monitoring technique by gel-phase ¹⁹F NMR spectrometry on cross-linked polystyrene derivatives proved to be of high value in four steps of the process by a comparison of data gathered from both a classic NMR spectrometer and elemental analysis.

4.3.1.3   Gel-phase ³¹P NMR

Gel-phase ³¹P NMR was used to monitor solid-phase oligonucleotide synthesis by Giralt and co-workers (Bardella et al. 1993). This method has several advantages. First, different phosphorous derivatives are involved in the key steps of the synthetic cycle in oligonucleotide synthesis. Second, its NMR sensitivity is much higher than that of ^{13}C (383/1 considering the natural abundance of the nuclei). Finally, the signals of the polymer do not interfere with the observed ^{31}P NMR resonances.

The half-height width of the ^{31}P NMR signal of resin-bound DMT-C^{bz}-P(Ar)-G^{ib} is 26 Hz. The ^{31}P NMR spin-lattice relaxation time (T_1) of this resin-bound compound measured by using the inversion recovery method is 1.006 ± 0.013 s. Considered together, linewidth and T_1 values provide a clear indication that the considerable degree of swelling of the dinucleotide-resin in CDCl$_3$ provides a highly flexible microenvironment for the phosphorous atoms and, therefore, of the reactive sites. By showing several reaction examples, authors have pointed out that ^{31}P NMR is a useful technique for monitoring solid-phase oligonucleotide synthesis and is particularly interesting in methodological studies in order to ensure quantitative yields or to determine the extent of a side reaction on the resin. The use of ^{31}P nuclei for NMR provides great sensitivity to the analysis, especially considering the size of this kind of molecule, and takes advantage of the fact that the solid support is transparent. In addition, the great difference between the chemical shifts of P^{III} and P^V makes gel-phase ^{31}P NMR especially useful as a nondestructive method for monitoring the synthesis of oligonucleotide using the phosphorous-triester approach.

Gel-phase ^{31}P NMR was used to monitor a synthesis of the solid-supported α, β-unsaturated amide using the reaction of phosphonoacetamide esters as Horner-Wadsworth-Emmons (HWE) reagents with a variety of aldehydes (Johnson and Zhang 1995). The reaction progress was monitored by the gel-phase ^{31}P NMR of the resin suspended in acetonitrile. Gel-phase ^{31}P NMR shows the resonance of the starting resin-bound diethylphosphonoacetamide as a narrow multiplet at δ22. After the reaction is complete, the δ22 peaks have disappeared, and a new broad resonance for diethylphosphate has appeared near 80.

4.3.1.4   Gel-phase ^{15}N NMR

Incorporation of an isotopically enriched ^{15}N-Alanine into the synthetic scheme and the application of gel-phase ^{15}N NMR were used to monitor a synthetic pathway that provides functionalized piperazin-2-one ring systems (Swayze 1997). This technique provided a convenient label for following the progress of the reaction sequence, as the amino-acid-derived nitrogen is involved in the key-ring-forming reactions and, therefore, exhibits significant differences in chemical shift. The sensitivity obtained was sufficient to gauge the success or failure of the reaction, as byproducts were evident by the presence of multiple peaks when inappropriate reaction conditions were chosen.

4.3.1.5   Gel-phase deuterium NMR

Gel-phase deuterium NMR spectra were reported for a series of swollen Merrifield resins containing protected glycine oligomers (Ludwick et al. 1986), with the goal of delineating molecular-level interactions that can affect the desired reactivity of these materials. Figure 4.6 shows that the static signal increases as the chain length increases. Heating samples (gly-d_2)$_7$ and (gly-d_2)$_9$ to 111°C resulted in significant decreases in the static portions of the spectra (Figure 4.7). Results were consistent with a model in which partial aggregation of the glycine oligomers occurs after the pendant chain reaches a critical length (n >5). Lengths greater than this correspond to the overlap necessary to form at least one helix repeat of the polyglycine II structure. The polystyrene matrix is, therefore, concomitantly immobilized, presumably caused by the additional effective cross-links caused by the aggregation. The mechanism shown in this work

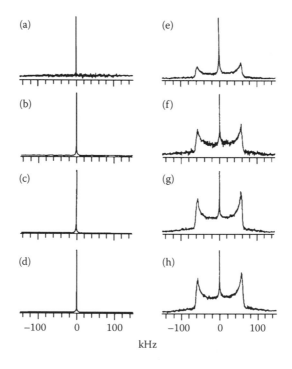

Figure 4.6 Quadruple echo deuterium NMR spectra of DMF swollen materials for (a) the polystyrene matrix and for solid-phase peptide samples (b) I (gly-d_2), (c) II ((gly-d_2)$_3$), (d) III ((gly-d_2)$_5$), (e) IV ((gly-d_2)$_7$), (f) V((gly-d_2)$_8$), (g) VI ((gly-d_2)$_8$) gly, and (h) VII ((gly-d_2)$_9$). The spectra were obtained with a recycle delay time of 2 s. (Reproduced with permission from *J. Am. Chem. Soc.* 1986, 108, 6493–6496. Copyright© 1986, American Chemical Society.)

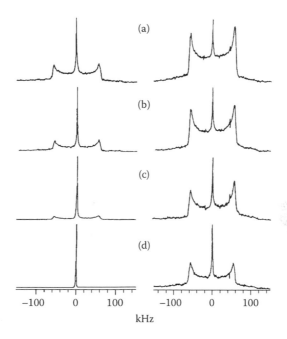

Figure 4.7 Quadrupole echo deuterium NMR spectra of DMF swollen sample IV (left column) and DMF swollen sample VII (right column) obtained at (a) 23, (b) 50, (c) 82, and (d) 111°C. (Reproduced with permission from *J. Am. Chem. Soc.* 1986, 108, 6493–6496. Copyright© 1986, American Chemical Society.)

agrees with other works presented in Section 2.3.4. and explains some difficult and inefficient synthesis of particular peptides.

4.3.2 Magic angle spinning (MAS) NMR

High-resolution magic angle spinning (HR MAS) NMR is a tool for characterizing organic reactions on solid support. Solid-supported chemical transformations are most commonly monitored by cleaving the reaction products from the polymer, followed by the application of standard solution-phase analytical methods. A significant line broadening in standard solution-phase NMR methods can be significantly reduced by the application of magic angle spinning. Besides, the method also allows increased sensitivity relative to gel-phase NMR, with the practical outcome being reduced acquisition times (Shapiro and Gounarides 2000).

HR MAS NMR is used to do detailed structure analysis and to monitor reactions. Application of a diffusion filter helps suppress solvent signals and dipolar recoupling. Using HR MAS NMR, resin-supported molecules can be rapidly analyzed in detail without any modifications of the sample.

Intrinsic linewidths of the NMR signals are especially substantial in ¹H spectra, which used to make it unrealistic to suppress purely solid-state NMR linewidths that can reach tens of kHz, even with the most powerful and sophisticated techniques such as CRAMPS. In practice, an HR MAS NMR experiment requires mainly hardware capable of spinning the cylindrical rotor that contains the swollen sample at the magic angle at KHz speed. Its probe is designed with magnetically matched materials that have to be pasted as symmetrically as possible around the solenoid coil. The typical sample amount of the resin for HR MAS experiments is 5–10 mg (Schroder 2003).

Solvent swelling is needed to make the active functional groups accessible to reagent in synthesis. The solvent motion in the resin network leads to chain expansion observable on a macroscopic scale as the swelling of the bead and is accompanied by an enhanced local mobility of the attached molecules at the microscopic level. As a consequence of the enhanced molecular mobility, the various factors that govern the linewidth of the NMR spectrum will be motionally averaged. These factors include the homonuclear dipolar interaction that can broaden proton signals in a solid-state sample. However, this motional averaging is incomplete because of the anisotropic environment of the solvent molecules in the resin. In addition to these magnetic-susceptibility interface differences within the sample, magnetic-susceptibility interfaces inherent in the construction of the probe are another factor for spectral linebroadening. It has long been recognized that the magnetic-susceptibility-induced linebroadenings in heterogeneous samples can be eliminated by spinning the sample at the magic angle (54.7° from Z) (Drain 1962; Doskocilova and Schneider 1970) at a spinning rate higher than the nonspinning residual linewidth. MAS has provided substantial line narrowing in a one-dimensional ¹³C and ¹H NMR spectrum of solvent-swollen polymer gels with conventional MAS probes. MAS removes the magnetic-susceptibility broadenings caused by the difference in magnetic susceptibility between the resin and the pure solvent, combined with the irregular shape of the solvent/resin interface. MAS is also advantageous for suppressing the lines of the more rigid supporting resin while obtaining more liquid-like linewidth for the more mobile attached compounds. However, the contribution of magnetic-susceptibility discontinuities within the materials used for the probe construction will not be eliminated by MAS, and can be removed only by a careful probe design. Very narrow ¹H linewidths (1–8 Hz) have been obtained with a probe called the Nanonmr™ probe, which combines the use of both MAS and "magnetic-susceptibility-matched" materials.

4.3.2.1   1D NMR spectroscopy of resin samples

Three possible high-resolution magic angle spinning (HR MAS) NMR experiments to quantitatively monitor a solid-phase supported

Horner-Emmons reaction are presented (Warrass and Lippens 2000). In the first experiment, the solid-phase reaction is followed in deuterated solvent directly in the NMR rotor. The second quantification is done by reconditioning a few milligrams of resin from an undefined reaction vessel by washing, drying, and reswelling in deuterated solvent, and then evaluating the amount of resin-bound structures by comparing to an external standard. The third experiment represents the first analytical quantification of resin-bound structures without any sample preparation, except for the transfer of resin-solvent suspension (a large excess of reagents in protonated dimethylformamide) from the reaction vessel to the NMR rotor.

These three experiments demonstrate how HR MAS NMR can be used to quantitatively monitor SPOC; the third method turned out to be best. So it is believed that HR MAS NMR methods will enter the solid-phase chemistry laboratory as an indispensable tool, similar to liquid NMR spectroscopy for organic chemistry in homogeneous phase.

The quality of NMR spectra obtained for resin-bound N-Fmoc-L-aspartic acid using conventional high-resolution, MAS, and high-resolution MAS (Nanonmr™) probes was compared by Keifer and co-workers (1996). Because the primary criterion of quality in NMR spectra for resin samples is the resonance linewidth, all possible factors that may affect linewidth were investigated and discussed. Factors that affect the achievable linewidth include the swelling of the resin with solvent, which increases molecular mobility; the microscopic discontinuities of magnetic susceptibility within the resin, which can be removed by MAS; the mobility of the nuclei at the end of the tether, which is controlled by both the length of the tether and the solvation; the spectral-resolution limitation of the probe; and within the more rigid region of the resin polymer, possible solid-state effects. Magnetic susceptibility differences in the sample are the dominant line-broadening mechanism for the resonances of the mobile nuclei within the resin. Magnetic-susceptibility mismatches within the sample are removed by spinning the sample about the magic angle, while magnetic susceptibility mismatches within the probe are not. The largest difference was seen in ^{1}H spectra, which became significantly narrower with MAS and using the ^{1}H Nano probe. The three main findings from this study are that 1) ^{1}H spectra must be obtained in a Nanoprobe, whereas ^{13}C spectra could also be obtained in either a Nanoprobe or a CP/MAS probe; 2) a Nanoprobe will always generate higher resolution spectra than a CP/MAS probe; 3) the Nanoprobe will be the most efficient choice for small quantities of sample while a CP/MAS probe may provide higher S/N spectra if large quantities of sample are available.

Lacey and co-workers (2001) introduced the use of a custom-built NMR microprobe to generate high-resolution ^{1}H NMR spectra from the cleaved product of individual 160-μm Tentagel beads. By injecting a dissolved sample sandwiched between an immiscible, perfluorinated organic liquid directly into a capillary microcoil flow probe, NMR spectra of the product cleaved from single beads were acquired in just ^1H of NMR spectrometer time. Using the relative intensity of the DMSO-d5H versus the analyte signals in a fully relaxed CPMG spectrum, the amount of product cleaved from a single bead was determined to be 540±170 pmol in one of the samples. The flow probe configuration facilitates coupling to other analytical systems such as UV-Vis spectroscopy, FT-IR, and MS, in an online manner. The cleaved product from single beads can be subjected to a plethora of characterization techniques to complete the structural elucidation of single-bead products (Lacey et al. 2001).

Fitch and co-workers (1994) reported the application of the high-resolution ^1H NMR in SPOS. The conventional 500 MHz NMR spectrum of product 1 (Diagram 4.1) in DMSO-d6 was obtained using either a conventional 5-mm liquid probe (top spectrum in Figure 4.8) or by spinning at 2000 Hz about the magic angle in a Varian Nanonmr probe (bottom spectrum in Figure 4.8).

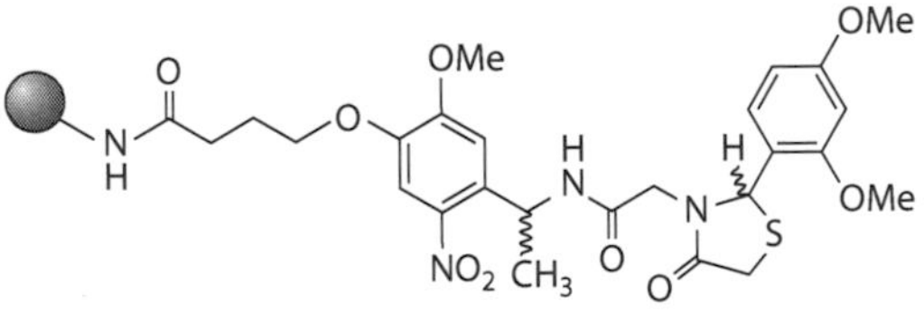

3

Diagram 4.1

A compound library is synthesized with solid organic synthesis or other methods. Structural validation and quantification for every member of the library were achieved by evaluation of ^1H NMR and HPLC, respectively. The ^1H NMR data were obtained by using automated delivery of DMSO solutions of every library member from a 96-well plate to a flow-probe-equipped NMR spectrometer. HPLC data were used for determining the extent of conversion of malonamic/malonate ester 2 to the products by an external standard method. Summary information from the ^1H NMR and HPLC data is viewed as plate diagrams for analysis of the final library (Hamper et al. 1999).

The application of the analytical construct methodology to the high-throughput analysis of samples derived from single beads. A typical analytical construct contains a linker that allows the release of substrates in the conventional manner, as well as an orthogonal linker that is cleavable by a specific reagent in an additional analytical mode. These dual-linker analytical constructs have been designed to address problems of monitoring and analysis of reactions conducted on solid support and have been reported to reliably render all compounds visible to mass spectrometry,

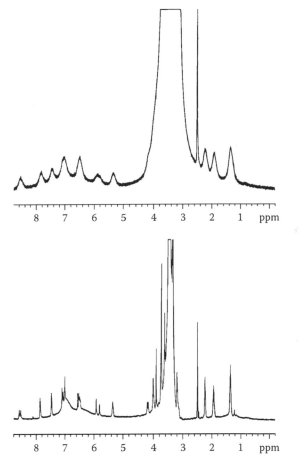

Figure 4.8 (Top) A 16-scan 500 MHz NMR spectrum of 3 (100 mg suspended in 600 μl of DMSO-d$_6$). The spectrum was obtained using a conventional 5-mm liquids probe. (Bottom) A 16-scan 500 MHz NMR spectrum of 3 (10 mg suspended in 30 μl of DMSO-d$_6$). The spectrum was obtained by spinning at 2000 Hz about the magic angle in a Varian Nano-NMR probe. (Reproduced with permission from *J. Org. Chem.* 1994, 59, 7955–7956. Copyright© 1994, American Chemical Society.)

even from a single bead. The single-bead ESI-MS data that were collected allowed for a direct comparison of the analysis conducted after conventional cleavage with the analysis using analytical construct cleavage (Lorthioir et al. 2001).

Although the PS-PEG resin has been proven to possess the best NMR properties with respect to linewidth and spectral purity, high-quality MAS NMR spectra of polystyrene-based resins have been obtained (Wehler and Westman 1996). Wehler and Westman showed that PS-based resins are well suitable for NMR analysis with a Nanoprobe.

The utility of the spin echo sequence (90°-τ–180° -τ) with a selected τ value, along with MAS at a speed of 20 kHz in suppressing signals from the PS matrix of the resin, has been reported (Garigipati et al. 1996). It was shown that by the appropriate choice of τ value, one can significantly minimize the effect of polystyrene peaks.

4.3.2.2   2D NMR *spectroscopy of resin samples*

MAS ^{1}H NMR spectra of resin-bound compounds, with the exception of TentaGel, are relatively broad and featureless. The additional complication of polymer peaks can be attenuated by using a spin echo sequence, although the residual peaks can still be problematic. The loss of coupling information relegates the interpretations to chemical shift information only, making peak assignment difficult. Coupling information can be reintroduced by performing a 2D J-resolved experiment. Usually, the 2D-J-resolved spectrum is "tilted" by rotating the data to align the chemical shift axis so that, upon projection of the data, a "proton-decoupled" proton NMR spectrum can be obtained. However, the untilted projection can yield 1D proton NMR spectra of resin-bound compounds that are of high quality. The top spectrum in Figure 4.9a is the spin-echo MAS NMR spectrum for DMF-swollen crude product of isoleucine on Wang resin obtained by Fmoc chemistry (Shapiro et al. 1997). The quality of the spectrum is typical of this derivatized 1% DVB cross-linked PS resin. In contrast, the bottom spectrum in Figure 4.9b shows the spectrum obtained for the same compound by projection onto the chemical shift axis of the untilted 2D-J-resolved data. The resonance patterns for the two methyl groups of isoleucine (inset) serve as a comparison of the differences in the resolution.

The reduction of inhomogeneous linebroadening in 2D MAS NMR spectra of molecules attached to swollen resin and the extraction of crucial J-coupling-constants information otherwise buried in the linewidth were achieved by using a pulse sequence for multiple quantum-filtered COSY or E.COSY with scaling in evolution period (Meissner et al. 1997). By swelling the resin and spinning it at the magic angle, linebroadening caused by the restricted molecular motions within a sample and the magnetic susceptibility are reduced. The resulting linewidth is still not narrow enough to reveal the J-coupling constant. Because the inhomogeneity is associated with the chemical shift terms of Hamiltonian, a solution is to scale down these terms relative to the J-coupling terms. This is done by inserting the element $kt_{1/2}$-π-$kt_{1/2}$ into the evolution period of any existing 2D correlation sequence as indicated for multiple quantum-filtered COSY (Piantini et al. 1982). In this way, the chemical shifts are scaled by a factor of $(k+1)^{-1}$ while the J-coupling is unaffected.

A complete NMR structure analysis of an eight-residue peptide, covalently bound to a single bead of a poly(ethyleneglycol)-based (POEPOP

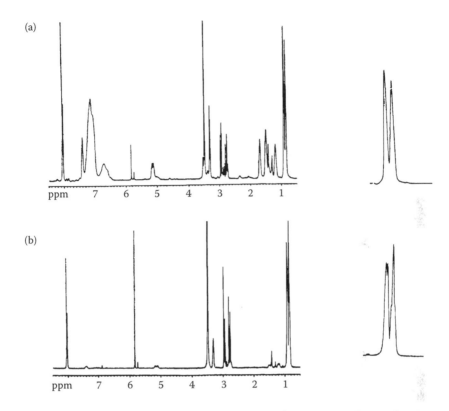

Figure 4.9 (a) Spin echo MAS ¹H NMR spectrum for DMF swollen isoleucine on Wang resin. Inset shows the two methyl resonances at δ = 0.95 and 0.89. (b) Projection of the homonuclear 2D-J-resolved MAS of DMF swelled isoleucine on Wang resin. Inset shows the methyl resonances at δ = 0.95 and 0.89. (Reproduced with permission from *Tetra. Lett.* 1997, 38, 1333–1386. Copyright© 1997, Elsevier Science Ltd.).

1500) resin, is described. Well-resolved 1D and 2D ¹H NMR spectra were obtained using magic angle spinning (MAS) Nanoprobe NMR spectroscopy at 500 MHz. In the experiment, it was found that about 1.6 nmol of peptide was the minimum amount needed to obtain a high-resolution 1D ¹H NMR spectrum in less than 20 minutes, while about 6 nmol of peptide on a single bead is required for a complete structure elucidation by 2D ¹H NMR (Gotfredsen et al. 2000).

The acquisition of the MAS ¹³C-¹H-correlated NMR of resin-bound compounds with a standard solid-state probe has been reported (Anderson et al. 1995a). The MAS ¹³C NMR spectrum of a norbornane-2-carboxylic acid bound to Wang resin showed carbonyl resonance at 174.9, 173.8 ppm separately, which was not observed in the gel-phase NMR spectrum of the same sample. The ratio of exo/endo epimers was semi-quantitatively determined as 60:40, based on the peak intensities in the MAS ^{13}C spectrum. The resonances of tether carbon atoms was also observed, and these signals provide a straightforward means of determining the extent of reaction, since an internal reference is observed in the resonances of the tether. The linewidth of 26 Hz obtained in these experiments allows structural information to be obtained by evaluation of CH-correlated data. The MAS CH-correlated spectrum of these compounds (shown in Figure 4.10a) and the solution-state spectrum of norbornane-2-carboxylic acid (Figure 4.10b) are shown for comparison. This technique promises to be useful not only in the determination of the success or failure of a chemical transformation, but also in the structural verification of molecules produced by SPOS, as well as detection of diastereomers and side products.

The total correlation spectroscopy (TOCSY) NMR spectrum and the heteronuclear multiple quantum correlation (HMQC) spectrum of Wang-Fmoc-lysine-Boc were obtained. Correlations of the lysine portion of the molecule are readily observed. The connectivity through the chain can be traced from the NH group to the α-proton through the ε protons and in the reverse direction. The chemical shift assignment follows directly from consideration of the TOCSY data and from expected ^{1}H and carbon ^{13}C chemical shift relationships (Anderson et al. 1995b).

It would be advantageous if the enhanced resolution afforded by the projection of the 2D-J-resolved data can be retained while spin system connectivity is also obtained. Although correlation spectroscopy (COSY) and TOCSY data allow for the connectivity to be obtained, the projection onto the chemical shift axis does not lead to enhanced resolution. In spin-echo correlated spectroscopy (SECSY), both spin connectivity and resolution enhancement can be obtained (Chin et al. 1998). The apparent increase in resolution and the ability to make spin-spin connectivity were shown by Wang resin-bound compounds. A major problem associated with 2D NMR data obtained using MAS is that artifacts and T_1 noise are abundant, leading to potential confusion in data interpretation. By using the gradient MAS probe, this problem is alleviated.

As in high-resolution liquid-state studies, the gradient is essential for removing T_2 noise and is extremely helpful in coherence pathway selection. A series of heteronuclear correlation experiments, heteronuclear multiple quantum correlation (HMQC), and heteronuclear multiple bond correlation (HMBC) were carried out, and the level of artifact in MAS spectra of Wang-bound N-Fmoc-N-Boc-L-Lysine swollen with CDCl$_3$ was suppressed by using gradients (Maas et al. 1996). It is anticipated that gradient methods will become as common in MAS as they are in high-resolution liquid-state measurement. In fact, the need for gradient is more pronounced in the MAS case because the probe geometry and sample spinning build into the measurement an unavoidable modulation of the coupling of the spins to the receiver and, hence, pronounced artifacts.

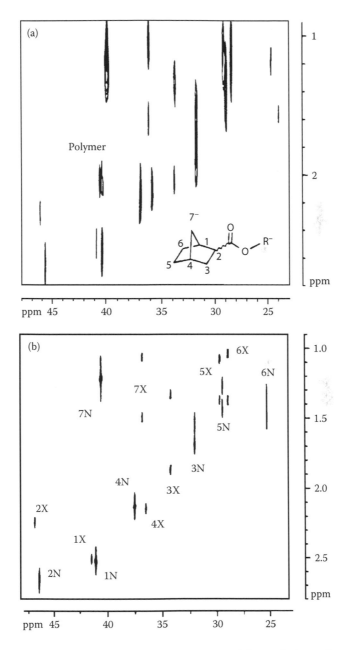

Figure 4.10 CH-correlated MAS NMR spectrum of the resin-bound compound (a) swollen in benzene, where R=Wang resin. (b) Solution CH-correlated spectrum for a mixture of epimers of norbornane-2-carboxylic acid. Assignments are indicated for endo (N) and exo (X) epimers. (Reproduced with permission from *J. Org. Chem.* 1995, 60, 2650–2651. Copyright© 1994, American Chemical Society.)

A 2D ^{1}H NOE spectrum of a hexapeptide GPFRAF bound to amino PS-PEG resin was obtained (Jelinek et al. 1997). Short peptides in solution usually do not yield any NOE cross-peaks because of their fast molecular motion and conformational flexibility. In this 2D ^{1}H NOE spectroscopic study, a relative large number of NOE cross peaks were found. This suggests that a significant part of the population of the peptide on beads is folded into a defined stable conformation with less molecular motion than for the same peptide in solution.

A three-step SPOS involving a Heck reaction was monitored by a series of 2D MAS NMR measurements (Pop et al. 1996). All intermediate products were characterized on resin. The work demonstrated that the goal of reaction characterization and optimization *in situ* can be obtained in an entirely satisfying manner using conventional MAS NMR, although each 2D NMR experiment required 3 hours to complete.

4.3.2.3   NMR on multipin crowns

SPOS uses resin beads as the primary format for synthesis. An alternate format is the multipin crown system where polystyrene is grafted onto a base polymer unit called a "crown," and arrays of these units are held on the ends of pins in a 96-array format for parallel synthesis. The first MAS NMR spectrum of crowns was reported (Chin et al. 1997). Spectra from a product of the Wittig reaction was recorded and clearly showed the disappearance of the peak from the starting material (aldehyde $\delta=10.17$) and the formation of a peak from the product (olefin, $\delta=6.78$, 7.72). The procedure is similar to that for resin beads. This work opened another possibility for effectively monitoring reactions carried out on multipin crowns.

4.3.3   Other NMR methods

4.3.3.1   Flow NMR spectroscopy in drug discovery

Flow NMR spectroscopy techniques are becoming increasingly utilized in drug discovery and development. LC-NMR has become a routine method for resolving and identifying mixtured components because of its broad applications in natural products, biochemistry, and drug metabolism and toxicology studies. The concomitant development of robust automated data analysis has facilitated the interpretation of the vast amount of NMR data generated. Generally mentioned are LC-NMR, direct-injection NMR, and automated data analysis tools (Stockman et al. 2000).

4.3.3.2   Flow injection analysis NMR (FIA-NMR)

Flow injection analysis NMR (FIA-NMR), as a novel flow NMR technique, complements LC-NMR and direct injection NMR (DI-NMR). Liquid chromatography NMR and direct injection NMR are used together. FIA-NMR is shown to be useful as an analytical technique, especially for repetitive analyses, and may also prove useful in the analysis of combinatorial chemistry libraries. All this is proved by the following experiment:

A Varian UNITY-INOVA 500 MHz spectrometer is equipped with a standard LC-NMR hardware and software package. The HPLC injector port was replumbed to be directly connected to the NMR probe with 10 ft of 0.010 in-capillary tubing. Injector loops having volumes of 20 to 200 µL were used. All spectra were referenced to the acetonitrile resonance at 1.95 ppm.

FIA-NMR can handle a wide variety of sample volumes. Its plumbing scheme facilitates the addition of sample filtration. It offers more control over the amount of probe rinsing and hence the amount of sample-to-sample carryover. Compared to DI-NMR, FIA-NMR has these three advantages. But its disadvantages are also distinct. For example, sample recovery is more difficult and it is almost impossible to recover samples without dilution (Keifer 2003).

4.3.3.3   High-throughput NMR analysis using a multiple coil flow probe

An automated method for high-throughput nuclear magnetic resonance spectroscopy has been developed using a four-coil Multiplex NMR probe. The probe is constructed with solenoidal microcoils optimized for detection of small volume, mass-limited samples and a flow-through design. Four samples can be simultaneously injected into the Multiplex probe with a robotics liquid handler and then analyzed in rapid succession using a selective excitation experiment. Because of the simultaneous injection of four samples and the reduced analysis time with rapid selective excitation, the analysis rate achieved thus is as high as 1 sample/34 s for 1D ^{1}H NMR (Macnaughtan et al. 2003).

4.3.3.4   NMR detection with multiple solenoidal microcoils for continuous-flow capillary electrophoresis

As the separation channel diameter becomes smaller, the sample loading capacity of capillary electrophoresis (CE) also decreases. However, nuclear magnetic resonance (NMR), which provides critical structural information, currently requires relatively large sample quantities. A new method for avoiding microcoil NMR spectral degradation during continuous-flow CE is demonstrated using a unique multiple solenoidal coil NMR probe. The electrophoretic flow from a single separation capillary is split into multiple outlets, each possessing its own NMR detection coil. While the CE electrophoretic flow is directed through one outlet, stopped flow, high-resolution NMR spectra are obtained from the coil at the other outlet (Wolters et al. 2002).

4.3.3.5 Quantitative measurements in continuous-flow HPLC/NMR

Two methods are introduced for quantitative determination of compounds in continuous-flow HPLC/NMR. The first method uses an internal standard (caffeine) of known concentration directly mixed into the mobile phase, while in the second method a known amount of internal standard is injected onto the column during the chromatographic run. The second method was validated using several nitroaromatic compounds and explosives. Deviations between the injected and calculated amounts of analytes are usually below 10%, while the relative standard deviation ranges from 2% in the upper microgram range to 40% at the limit of detection. If an internal standard is added to the sample, quantification of complex mixtures of organic compounds by on-flow HPLC/NMR is possible (Godejohann and Preiss 1998).

4.3.3.6 Reaction monitoring in LPOS by ^{19}F NMR

^{19}F NMR has offered an opportunity for monitoring the reactions on support on both the spacer and the linker motifs. The resonances of different fluorous nuclei were usually well separated using F NMR analysis. Lakshmipathi and co-workers (2002) prepared different fluorous spacers, such as monofluorospacers, difluorospacers, and perfluorospacers, attached to soluble polymer support and set up different LPOS. Of the three types of spacers studied, they found perfluoro Wang linker to be the most efficient for the purpose, which could be easily anchored to and cleaved with the substrates. In addition, they found that the support could be recovered without loss in loading.

Experiments were performed to validate the quantitative NMR method for the analysis of organic chemicals. Examples of ^{1}H and ^{31}P NMR used for quantitative analysis of agricultural chemicals and a method for characterization in analytical standards are presented. The level of the major chemical ingredient can be determined with an accuracy and precision significantly better than 1%, and impurities may be quantified at the 0.1% level or below. QNMR can rival chromatographic runs in sensitivity, speed, precision, and accuracy while avoiding the need to match reference standards for each analyte. The QNMR method is considered universal and highly specific (Maniara et al. 1998). As a superior tool, ^{19}F NMR has many positive characteristics. ^{19}F NMR is excellent for the generation of an encoding method for single beads in combinatorial libraries produced by the mix-and-split synthetic method. A fluorine-encoding method has been used for the first position monomers in mix-and-split libraries for the purpose of assistance to mass spectrometry in deconvolution of library hits. Both mass spectral analysis of active bead and on-bead analysis of a first-position nonreleasable code are utilized to reduce the possible structures to one in every case (Hochlowski et al. 1999).

4.4 Summary

Although most MS and NMR applications in the reaction optimization stage are still following the "cleave and analyze" protocol, on-resin applications have greatly increased. MS analysis always requires the cleavage before analysis. Two methods have accelerated the cleavage speed. One uses light and a photocleavable linker to release the bound compound. The other method uses TFA vapor with an acid labile linker. These methods at least have gotten rid of the time-consuming problem of the "cleave and analyze" protocol.

There is a long history of using gel-phase NMR for monitoring reactions on resin. On-resin NMR can be routinely obtained, and the most useful nuclei are ^{13}C, ^{19}F, ^{31}P, and ^{15}N. However, the resolution and the spectral quality have not achieved the same level as those of solution samples. A question is whether we need the same resolution to accomplish our goal. For most SPOS optimization projects, the goal is to confirm the formation of a desired product, not elucidate an unknown structure. In many instances, this can be achieved without a solution-like quality spectrum. The gel-phase sampling method is a convenient way to record an NMR spectrum. Although tens to a hundred milligrams of resin are required for analysis, the method is still useful because the sample amount is not limited in reaction optimization work and the NMR sample can be reused for synthesis after analysis. Finally, the most important aspect is that the probe and instrumentation required for acquiring gel-phase NMR are widely available in the chemistry laboratory.

MS/NMR is used as a structure-based approach for discovering protein ligands and for drug design by coupling size-exclusion chromatography, mass spectrometry, and nuclear magnetic resonance spectroscopy. A protocol is described for rapidly screening small organic molecules for their ability to bind a target protein while obtaining structure-related information as part of a structure-based drug discovery and design program. Size exclusion gel chromatography, mass spectrometry, and NMR are combined together for high-throughput screening. Size exclusion gel chromatography provides high-speed separation of a protein-ligand complex from free ligands. The spin column eluent is then analyzed under denaturing conditions by electro-spray ionization mass spectrometry (EI MS) for the presence of small molecular weight compounds formerly bound to the protein. A 2D ^{1}H-^{15}N HSQC NMR spectrum is then used to identify the structure of the molecule. And the structure is finally determined by NMR, X-ray, and modeling. This is all for designing a chemical library from hits (Moy et al. 2001).

A rapid, microcomputer-based search system for binary-coded spectra is demonstrated for large infrared, ^{13}C-NMR, and mass spectral data containing 918725, 2229, and 30476 spectra, respectively. The search system exhibited is entirely "free-standing" and does not depend upon the availability of any other data-processing requirements for pre- or postsearch processing.

The unknown spectrum is compared to the spectra of a library of known compounds, and a list of candidate compounds is given as the result of the research. If the unknown compound is contained in the library, a match should be reported between the unknown spectrum and its corresponding entry in the library, and thus the unknown compound would be identified.

With this medium, roughly 50% of the total search time is spent reading data from the tape into memory, so the search is moderately I/O bound (Alan et al. 1982).

MAS NMR and the high-sensitivity probes have greatly improved the resolution for ¹H spectra with a much smaller amount of sample. 2D NMR spectra are also becoming available. Although there is a long way to go before an unknown structure can be resolved on-resin routinely, the MAS technique can help solve various problems in the optimization efforts.

References

Alan, P., Uthman, J.P., Koontz, J., Hinderliter-Smith, W., Stephen, W., Charles, N.R. 1982. High-throughput microcomputer-based binary-coded search systems for infrared, carbon-13 nuclear magnetic resonance, and mass spectral data. *Anal. Chem.* 54, 1772.

Anderson, R.C., Jarema, M.A., Shapiro, M.J., Stokes, J.P., Ziliox, M. 1995a. Analytical techniques in combinatorial chemistry: MAS CH correlation in solvent-swollen resin. *J. Org. Chem.* 60, 2650.

Anderson, R.C., Stokes, J.P., Shapiro, M.J. 1995b. Structure determination in combinatorial chemistry: Utilization of magic angle spinning HMQC and TOCSY NMR spectra in structure determination of Wang-bound lysine. *Tetra. Lett.* 36, 5311.

Andrew, M.W., Dimuthu, A.J., Andrew, G.W., Jonathan, V.S. 2002. Multiple solenoidal microcoil NMR detection for continuous-flow CE. *Anal. Chem.* 74, 5550.

Bardella, F., Eritja, R., Pedroso, E., Giralt, E. 1993. Gel-phase ³¹P-NMR: A new analytical tool to evaluate solid phase oligonucleotide4 synthesis. *Bioorg. Med. Chem. Lett.* 3, 2793.

Bayer, E., Albert, K., Willisch, H., Rapp, W., Hemmasi, B. 1990. ¹³C NMR relaxation times of a tripeptide methyl ester and its polymer-bound analogues. *Macromolecules.* 23, 1937.

Blossey, E.C., Cannon, R.G. 1990. Synthesis, reactions, and ¹³C FT NMR spectroscopy of polymer-bound steroids. *J. Org. Chem.* 55, 4664.

Brummel, C.L., Lee, I.N.W., Zhou, Y., Benkovic, S.J., Winograd, N. 1994. A mass spectrometric solution to address the problem of combinatorial libraries. *Science* 264, 399.

Carrasco, M.R., Fitzgerald, M.C., Oda, Y., Kent, S.B.H. 1997. Direct monitoring of organic reactions on polymeric supports. *Tetra. Lett.* 38, 6331.

Chávez, D., Ochoa, A., Madrigal, D., Castillo, M., Espinoza, K., González, T., Vélez, E., Meléndez, J., García, J.D., Rivero, I. A. 2003. Method for analysis of polymer-supported organic compounds using mass spectrometry direct insertion. *J. Comb. Chem.* 5, 149.

Chin, J., Fell, B., Pochapsky, S., Shapiro, M.J., Wareing, J.R. 1998. 2D SECSY NMR for combinatorial chemistry. High-resolution MAS spectra for resin-bound molecules. *J. Org. Chem.* 63, 1309.

Chin, J., Fell, B.; Shapiro, M. J., Tomesch, J., Wareing, J.R., Bray, A.M. 1997. Magic angle spinning NMR for reaction monitoring and structure determination of molecules attached to multipin. *J. Org. Chem.* 62, 538.

DeWitt, S.H., Kiely, J.S., Stankovic, C.J., Schroeder, M.C., Reynolds, D.M., Ravia, M.R. 1993. "Diversomers": An approach to nonpeptide, nonoligomeric chemical diversity. *Proc. Natl. Acad. Sci. U.S.A.* 90, 6909.

Doskocilova, D., Schneider, B. 1970. Narrowing of proton NMR lines by magic angle rotation. *Chem. Phys. Lett.* 6, 381.

Drain, I.F. 1962. The broadening of magnetic resonance lines due to field inhomogeneities in powdered sample. *Proc. Phys. Soc.* 80, 1380.

Egner, B.J., Bradley, M. 1997. Analytical techniques for solid-phase organic and combinatorial synthesis. *Drug Disc. Today* 2, 102.

Egner, B.J., Cardno, M., Bradley, M. 1995a. Linkers for combinatorial chemistry and reaction analysis using solid phase *in situ* mass spectrometry. *J. Chem. Soc., Chem. Commun.* 21, 2163.

Egner, B.J., Langley, G.J. Bradley, M. 1995b. Solid phase chemistry: Direct monitoring by matrix-assisted laser desorption/ionization time of flight mass spectrometry: A tool for combinatorial chemistry. *J. Org. Chem.* 60, 2652.

Epton, R., Goddard, P., Ivin, K.J. 1980. Gel phase ^{13}C N. M. R. spectroscopy as an analytical method in solid (gel) phase peptide synthesis. *Polymer* 21, 1367.

Fitch, W.L., Detre, G., Holmes, C.P. 1994. High-resolution ^{1}H NMR in solid-phase organic synthesis. *J. Org. Chem.* 59, 7955.

Fitzgerald, M.C., Harris, K., Shevlin, C.G., Siuzdak, G. 1996. Direct characterization of solid phase resin-bound molecules by mass spectrometry. *Bioorg. and Med. Chem. Lett.* 6, 979.

Gallop, M.A., Fitch, W.L. 1997. New methods for analyzing compounds on polymetric supports. *Curr. Opin. Chem. Biol.* 1, 94.

Garigipati, R.S., Adams, B., Adams, J.L., Sarkar, S.K. 1996. Use of spin echo magic angle spinning 1H NMR in reaction monitoring in combinatorial organic synthesis. *J. Org. Chem.* 61, 2911.

Gibson, C., Sulyok, G.A.G., Hahn, D., Goodman, S.L., Hölzemann, G., Kessler, H. 2001. Nonpeptidic αγβ3 integrin antagonist libraries: On-bead screening and mass spectrometric identification without tagging. *Angew. Chem. Int. Ed.* 40, 165.

Giralt, E., Rizo, J., Pedroso, E. 1984. Application of gel-phase ^{13}C-NMR to monitor solid. *Tetrahedron* 40, 4141.

Godejohann, M., Preiss A. 1998. High-resolution studies of hyaluronic acid mixtures through capillary gel electrophoresis. *Anal. Chem.* 70, 590.

Gotfredsen, C.H., Grøtli, M., Willert, M., Meldal, M., Duus, J. 2000. Single-bead structure elucidation: Requirements for analysis of combinatorial solid-phase libraries by nanoprobe MAS-NMR spectroscopy. *J. Chem. Soc. Perkin Trans.* 1, 1167.

Hamper, B.C., Snyderman, D.M., Owen, T.J., Scates, A.M., Owsley, D.C., Kesselring, A.S., Chott, R.C. 1999. High-throughput ^{1}H NMR and HPLC characterization of a 96-member substituted methylene malonamic acid library. *J. Comb. Chem.* 1, 140.

Haskins, N., Hunter, D., Organ, A.J., Rahman, S.S., Thom, C. 1995. Combinatorial chemistry: Direct analysis of bead surface associated materials. *Rapid Commun. in Mass Spectr.* 9, 1437.

Hochlowski, J.E., Whittern, D.N., Sowin, T.J. 1999. Encoding of combinatorial chemistry libraries by fluorine-19 NMR. *J. Comb. Chem.* 1, 291.

Hourdin, M., Gouhier, G., Gautier, A., Condamine, E., Piettre, S. R. 2005. Development of cross-linked polystyrene-supported chiral amines featuring a fluorinated linker for gel-phase F-19 NMR spectrometry monitoring of reactions. *J. Comb. Chem.* 7, 285.

Hughes, I. 1998. Design of self-coded combinatorial libraries to facilitate direct analysis of ligands by mass spectrometry. *J. Med. Chem.* 41, 3804.

Jelinek, R., Valente, A.P., Valentine, K.G., Opella, S.J. 1997. Two-dimensional NMR spectroscopy of peptides on beads. *J. Magn. Reson.* 125, 185.

Johnson, C.R., Zhang, B. 1995. Solid phase synthesis of aldenes using the Horner-Wadsworth-Emmons reaction and monitoring by gel phase ^{31}P NMR. *Tetra. Lett.* 36, 9253.

Jones, A.J., Leznoff, C.C., Svirskaya, P.I. 1982. Characterization of organic substrates bound to cross-linked polystyrenes by ^{13}C NMR spectroscopy. *Org. Magn. Reson.* 18, 236.

Keifer, P.A. 1997. High-resolution NMR techniques for solid-phase synthesis and combinatorial chemistry. *Drug Disc. Today* 2, 468.

Keifer, P.A. 1998. New methods for obtaining high-resolution NMR spectra of solid-phase synthesis resins, natural products, and solution-state combinatorial chemistry libraries. *Drugs of the Future* 23, 301.

Keifer, P.A. 2003. Flow injection analysis NMR (FIA-NMR): A novel flow NMR technique that complements LC-NMR and direct injection NMR (DI-NMR). *Magn. Reson. Chem.* 41, 509.

Keifer, P.A., Baltusis, L., Rice, D.M., Tymiak, A.A., Shoolery, J.N. 1996. A comparison of NMR spectra obtained for solid-phase-synthesis resins using conventional high-resolution, magic-angle-spinning, and high-resolution magic-angle-spinning probes. *J. Magn. Reson. A.* 119, 65.

Lacey, M.E., Sweedler, J.V., Larive, C.K., Pipe, A.J., Farrant, R.D. 2001. ^{1}H NMR characterization of the product from single solid-phase resin beads using capillary NMR flow probes. *J. Magn. Reson.* 153, 215.

Lakshmipathi, P., Crevisy, C., Gree, R. 2002. Reaction monitoring in LPOS by ^{19}F NMR: Study of soluble polymer supports with fluorine in spacer or linker components of supports. *J. Comb. Chem.* 4, 612.

Look, G.C., Holmes, C.P., Chinn, J.P., Gallop, M.A. 1994. Methods for combinatorial organic synthesis: The use of fast ^{13}C NMR analysis for gel phase reaction monitoring. *J. Org. Chem.* 59, 7588.

Ludwick, A.G., Jelinski, L.W., Live, D., Kintanar, A., Dumais, J.J. 1986. Association of peptide chains during Merrifield soid-phase peptide synthesis: A deuterium NMR study. *J. Am. Chem. Soc.* 108, 6493.

Lorthioir O., Carr, R.A.E., Congreve, M.S., Geysen, M.H., Kay, C., Marshall, P., McKeown, S.C., Parr, N.J., Scicinski, J.J., Watson, S.P. 2001. Single bead characterization using analytical constructs: Application to quality control of libraries. *Anal. Chem.* 73, 963.

Maas, W.E., Laukien, F.H., Cory, D.G. 1996. Gradient, high resolution, magic angle sample spinning NMR. *J. Am. Chem. Soc.* 118, 13085.

Macnaughtan, M.A., Hou, T., Xu, J., Raftery, D. 2003. High-throughput nuclear magnetic resonance analysis using a multiple coil flow probe. *Anal. Chem.* 75, 5116.

Manatt, S.L., Amsden, C.F., Bettison, C.A., Frazer, W.T., Gudman, J.T., Lenk, B.E., Lubetich, J.F., McNelly, E.A., Smith, S.C., Templeton, D.J., Pinnell, R.P. 1980. A fluorine-19 NMR Approach for studying Merrifield solid-phase peptide syntheses. *Tetra. Lett.* 21, 1397.

Maniara, G., Rajamoorthi, K., Rajan, S., Stockton, G.W. 1998. Method performance and validation for quantitative analysis by ^{1}H and ^{31}P NMR spectroscopy. Applications to analytical standards and agricultural chemicals. *Anal. Chem.* 70, 4921.

Meissner, A., Bloch, P., Humpfer, E., Spraul, M., Sorensen, O.W.L. 1997. Reduction of inhomogeneous line broadening in two-dimensional high-resolution MAS NMR spectra of molecules attached to swelled resins in solid-phase synthesis. *J. Am. Chem. Soc.* 119, 1787.

Moy, F.J., Haraki, K., Mobilio, D., Walker, G., Powers, R., Tabei, K., Tong, H., Siegel, M.M. 2001. MS/NMR: A structure-based approach for discovering protein ligands and for drug design by coupling size exclusion chromatography, mass spectrometry, and nuclear magnetic resonance spectroscopy. *Anal. Chem.* 73, 571.

Pastor J.J., Lingard, I., Bhalay G., Bradley M. 2003. Ion-extraction ladder sequencing from bead-based libraries. *J. Comb. Chem.* 5, 85.

Piantini, U., Sorensen, O.W., Ernst, R.R. 1982. Multiple quantum filters for elucidating NMR coupling networks. *J. Am. Chem. Soc.* 104, 6800.

Pop, I.E., Dhalluin, C.F., Déprez, B.P., Melnyk, P.C., Lippens, G.M., Tartar, A.L. 1996. Monitoring of a three-step solid phase synthesis involving a Heck reaction using magic angle spinning NMR spectroscopy. *Tetrahedron* 52, 12209.

Raftery, D. 2004. High-throughput NMR spectroscopy. *Anal. Bioanal. Chem.* 378, 403.

Rapp, W. 1996. PEG grafted polystyrene tentacle polymers: Physical-chemical properties and application in chemical synthesis. In *Combinatorial Peptide and Nonpeptide Libraries: A Handbook.* G. Jung, ed. 425. Wiley-VCH, Weinheim.

Schroder, H. 2003. High resolution magic angle spinning NMR for analyzing small molecules attached to solid support. *Comb. Chem. High Through. Screen.* 6, 741.

Shapiro, M.J., Chin, J., Marti, R.E., Jarosinski, M.A. 1997. Enhanced resolution in MAS NMR for combinatorial chemistry. *Tetra. Lett.* 38, 1333.

Shapiro, M.J., Gounarides, J.S. 2000. High resolution MAS-NMR in combinatorial chemistry: Analytical methods for the combinatorial sciences. *Biotech. and Bioeng.* 71, 130.

Shapiro, M.J., Kumaravel, G., Petter, R.C., Beveridge, R. 1996. ^{19}F NMR monitoring of a S_NAr reaction on solid support. *Tetra. Lett.* 37, 4671.

Stockman, B.J. 2000. Flow NMR spectroscopy in drug discovery. *Curr. Opin. Drug Disc. and Devel.* 3, 269.

Svensson, A., Fex, T., Kihlberg, J. 1996. Use of ^{19}F NMR spectroscopy to evaluate reactions in slide phase organic synthesis. *Tetra. Lett.* 37, 7649.

Swayze, E.E. 1997. The solid phase synthesis of trisubstituted 1,4-diazabicyclo[4.3.0] nonan-2-one scaffolds: On-bead monitoring of heterocycle forming reactions using ^{15}N NMR. *Tetra. Lett.* 38, 8643.

Sweeney, M.C., Pei, D. 2003. An improved method for rapid sequencing of support-bound peptides by partial Edman degradation and mass spectrometry. *J. Comb. Chem.* 5, 218.

Szewczyk, J.W., Zuckerman, R.L., Bergman, R.G., Ellman, J.A. 2001. A mass spectrometric labeling strategy for high-throughput reaction evaluation and optimization: Exploring C-H activation. *Angew. Chem. Int. Ed. Engl.* 40, 216.

Todd, M.H., Abell, C. 2001. Novel chemical tagging method for combinatorial synthesis utilizing Suzuki chemistry and Fourier transform ion cyclotron resonance mass spectrometry. *J. Comb. Chem.* 3, 319.

Warrass, R., Lippens, G. 2000. Quantitative monitoring of solid phase organic reactions by high-resolution magic angle spinning NMR spectroscopy. *J. Org. Chem.* 65, 2946.

Wehler, T., Westman, J. 1996. Magic angle spinning NMR: A value tool for monitoring the progress of reactions in solid phase synthesis. *Tetra. Lett.* 37, 4771.

Yu, Z., Yu, X.C., Chu, Y.H. 1998. MALDI-MS determination of cyclic peptidomimetic sequences on single beads directed toward the generation of libraries. *Tetra. Lett.* 39, 1.

chapter five

Reaction optimization using spectrophotometric and other methods

5.1   Introduction

Analytical methods based on spectrophotometric and fluorometric methods, in addition to qualitative analysis, provide quantitative means for analyzing organic compounds bound to solid supports. Traditional "cleaving, purifying, and weighing" methods cannot be applied to combinatorial chemistry because there is not always enough cleaved compound for weighing. On solid support, quantitation based on the weight gain of the support after synthesis is not accurate because this weight increase is usually very small and the entrapped reagent and solvent will cause error. Only if the weight gain is significant relative to the polymer backbones may this method be used to estimate the yield. If the "cleave and analyze" method has to be adopted, there are more appropriate choices of methods for quantitation of minute amounts of samples, such as HPLC quantitative detectors (the chemiluminescence nitrogen detector, evaporative light-scattering detector, and charged aerosol detector) (Zhang et al. 2008), NMR, and other analytical methods combined with a standard.

An important advantage of solid-phase organic synthesis (SPOS) is that it circumvents tedious intermediate isolation and purification procedures, such as recrystallization, distillation, and chromatography. Ideally, a solid-phase reaction can be driven to completion by excess reagent without causing problems during purification. In reality, most organic reactions cannot easily go to completion. Unlike solution synthesis, the resin-bound synthetic intermediate in SPOS cannot be purified. The unreacted portion will accumulate until the end of the synthetic reactions and cause purity problems for the final product. Therefore, on-resin quantitative analysis of solid-supported compounds at intermediate stages is very advantageous in guiding the multistep synthesis (Kay et al. 2000; Gaginni et al. 2004).

5.2   Qualitative analysis of amines

Resin lacking free amino groups exposed to the solution of bromophenol blue (3′, 3″, 5′, 5″-tetrabromophenolsulfophthalein; see Diagram 5.1) becomes yellowish to yellowy orange (λ_{max} 429 nm). When free amino groups are present, the color turns deep blue (91,800 M^{-1}cm^{-1} at 600 nm). Note that most acid-base indicators do not change their color in the presence of free amino groups on resin. Bromocresol green and bromocresol purple also displayed satisfactory properties. This simple method (Krchnak et al. 1988a, b) gave comparable results with those using either picric acid (Diagram 5.1) or the ninhydrin method (Scheme 5.1).

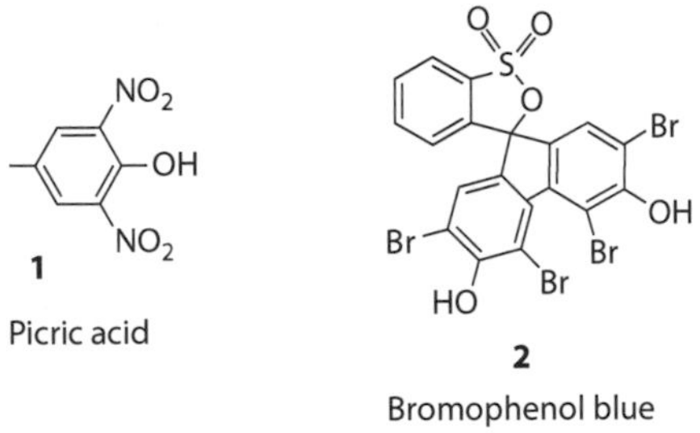

Picric acid **1**

Bromophenol blue **2**

Diagram 5.1

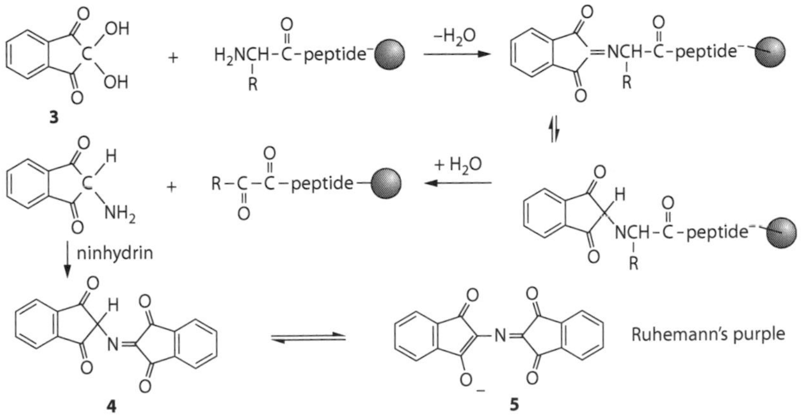

Scheme 5.1

The sensitivity of these monitoring methods brings about their own limitations:

1. The quality of the protected amino acids must be high. This high purity, in some cases, is very hard to achieve.
2. DMF used for the synthesis must be dimethylamine-free, and no base can be added during the coupling step.

3. The nature of the base used in the neutralization step is very important if the classical Merrifield resin is used. The residual chloromethyl groups form quaternary ammonium salts with the tertiary amine, and this leads to a blue coloring of the resin after the addition of the bromophenol blue.
4. Only a very low amount of monitoring dye can be used because it has the ability to decolorize itself by its leuko-form. Therefore, the amount of the indicator applied to the resin should be lower than the quantity of the free amino group that one wants to monitor.

A rapid microtest using 2,4,6-trinitrobenzenesulphonic acid (TNBS) has been developed (Hancock and Battersby 1976) to detect incomplete coupling reactions in solid-phase peptide synthesis. This new test detects amino groups at a level of 0.003 mmol/g.

One gram of resin reacts with one drop of 1% (w/v) solution of TNBS in DMF for 10 min. After centrifuging the mixture for five minutes to remove the supernatant, the resin appears red-orange under the microscope.

A qualitative test is described (Christensen 1979) that can be carried out, as some other reagents, in the presence of the reaction mixture. At intervals, a resin sample is removed from the reactor, and the chloranil test is carried out directly (Scheme 5.2). If free amino groups are present, a green to blue color develops on the resin beads in less than five minutes. The test can also be used when the N-terminal amino acid residue is proline. The sensitivity for detection of N-terminal proline is in the range of 0.002–0.005 mmol/g resin; for primary amino groups, it is in the range of 0.005–0.008 mmol/g when approximately 1 mg of peptide resin is assayed. Several peptide syntheses have been carried out and monitored by the chloranil test, indicating that the method is generally applicable. However, when histidine is present, the test result may be blurred. In a few cases, a discrepancy between the chloranil test and the ninhydrin test was observed.

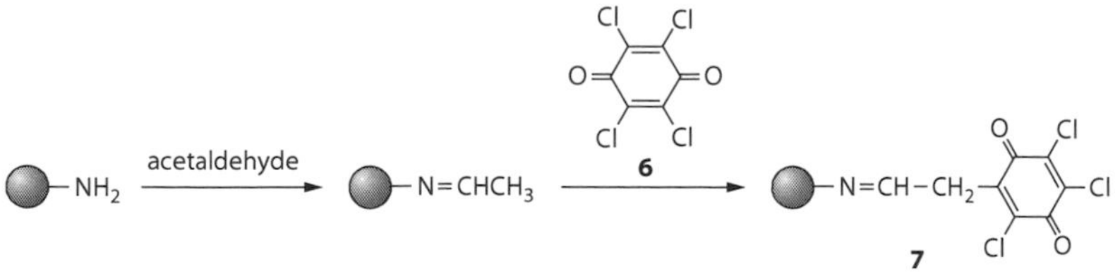

Scheme 5.2

One procedure (Vojkovsky 1995) uses chloranil in the presence of acetaldehyde to detect secondary amines. Except for N (p-tolyl) glycine, all secondary amino groups examined were detected reliably in a substitution range down to 0.0028 mmol/g. Protected imidazole in His (Trt) did not interfere with the detection method.

A method for the detection of aromatic amines on the solid support using chloranil was also reported (Mařík et al. 2003). The mixture of the resin with a 2% solution of chloranil in DMF was heated to 100°C for 5 min, with the resin turning red in the presence of a primary aromatic amino group. The chloranil reaction is specific to primary aromatic amines and can detect as little as less than 5 µmol/g of primary aromatic amines attached to the resin.

Proline is known to form a blue color with isatin (Scheme 5.3). The isatin test (Kaiser et al. 1980) made it possible to detect incomplete couplings to proline in cases when the ninhydrin test was so weak that it could be interpreted as negative. The positive isatin test became negative when the coupling reaction was repeated or when the reaction time was prolonged. Ala, Arg, Asp, Asn, Cys, Glu, Gln, His, Leu, Met, Thr, Phe, and Val all tested negative. Only with terminal prolyl residues did the color of the beads become blue. In addition to isatin, 4-N, N′-dimethylaminoazobenzene-4′-isothiocyanate (DABITC), and malachite green isothiocyanate (MGI) have been used in qualitative analysis of resin-bound amines (Shah et al. 1997).

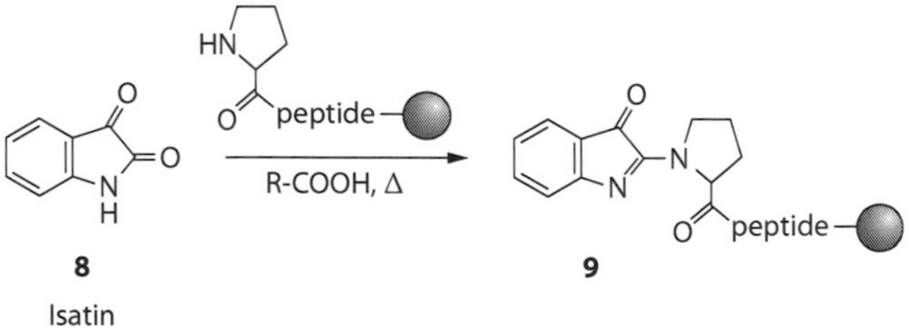

8

Isatin

Scheme 5.3

The fluorescamine test (Scheme 5.4) for the rapid detection of trace amounts of uncoupled products from solid-phase peptide synthesis has been reported (Felix and Jimenez 1973). Because the test is carried out under mild conditions, certain side reactions are circumvented. The fluorophore resins are easily viewed under long-wave ultraviolet light, are stable at room temperature, and may be used for quantitative evaluation.

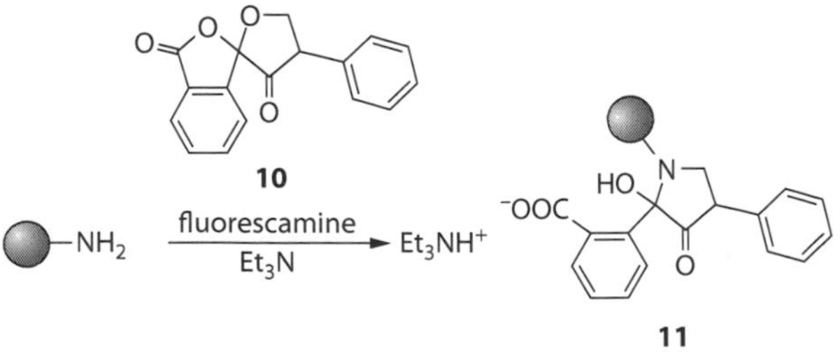

10

Scheme 5.4

Another highly sensitive fluorometric method (Garden and Tometsko 1972) for determining free amine groups on the Merrifield peptide resin is based on the reaction with dansylchloride (Scheme 5.5). The method is capable of quantitatively analyzing amine groups at a 0.2 μmol/g level, thus about ten-fold more sensitive than color test methods. The method was used to follow the kinetics of amino acid incorporation into resin-bound peptides.

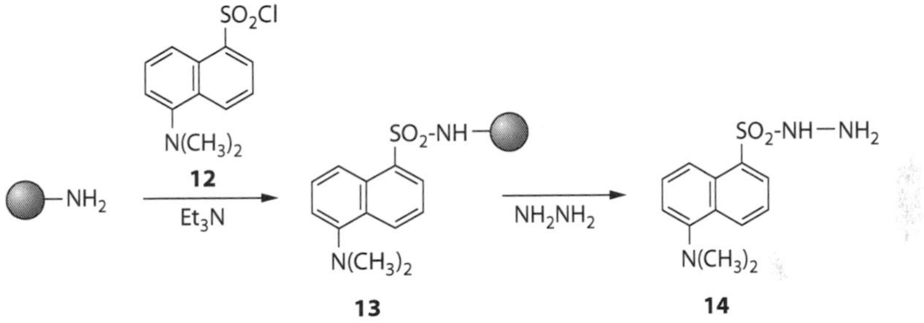

Scheme 5.5

Recently, a new sensitive and reliable colorimetric test (Clarerhout et al. 2008) was developed for the visual detection of resin-bound primary and secondary amines (Diagram 5.2), based on the formation of "on-bead-generated" colored stable intermediate azadiene between amines and 1-methyl-2-(4′-nitrophenyl)-imidazo[1,2-a]pyrimidinium perchlorate (DESC) (Scheme 5.6). The advantage of DESC exceeding the previously described solid-phase colorimetric amine tests is the high sensitivity for the detection of secondary and primary amines in a reliable and easy way (up to 2 mol % (9 μmol/g) of resin-bound free primary and secondary amine content). Another advantage of DESC is that the protocols for detection of primary and secondary amines using DESC are essentially different, enabling the selective detection of primary amines in the presence of secondary amines because the addition of DIEA (20% in DMF) is needed for the detection of secondary amines.

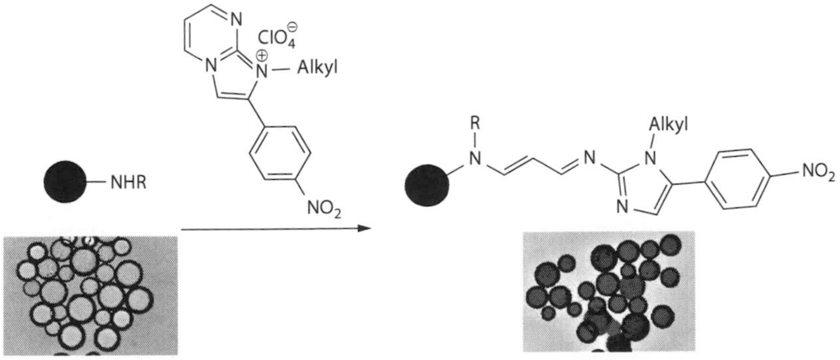

Diagram 5.2

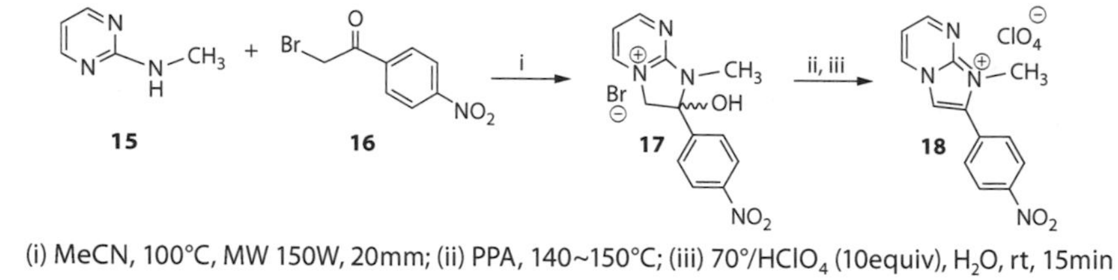

(i) MeCN, 100°C, MW 150W, 20mm; (ii) PPA, 140~150°C; (iii) 70°/HClO₄ (10equiv), H₂O, rt, 15min

Scheme 5.6

A new concept for solid-phase synthesis reaction monitoring, that of "self-indicating" resins, was developed using bromophenol blue (see Diagram 5.1) (Cho et al. 2003), in which resin-bound indicators act as integral sensors for the chemistry being carried out directly on the sensor bead itself. It allowed direct *in situ* monitoring of the chemistry being undertaken. Bromophenol blue was chosen as an indicator for its conspicuous color change from yellow to dark blue. To allow immobilization onto the resin without loss of conjugation, a carboxylic acid group was introduced onto bromophenol blue via a Suzuki cross-coupling reaction of bromophenol blue with 4-carboxyphenylboronic acid, allowing coupling to a number of solid supports. The absorption spectra compared well with that of bromophenol blue. λ_{max} slightly increased from 603 to 610 nm, while ε_{max} decreased from 110,000 to 73,000. The resin was successfully used as a "sensor" for monitoring solid-phase peptide synthesis, and applied in *in situ* reaction monitoring during the synthesis of a library of ureas in a highly successful way.

5.3   *Quantitative analysis of amines*

The "Kaiser Test" (Kaiser et al.; 1970, Scheme 5.1) is a rapid and sensitive method that determines if the formation of a peptide bond can reach 99.9% completion. This test is based on the reaction of ninhydrin reagents with resin-bound amines. The ninhydrin reagent described by Troll and Canan (1953) was found to be the most suitable for this purpose. In this reaction, a blue chromophore is formed. The absorbance of the violet-colored product, known as Ruhemann's purple, was observed at 570 nm with an average extinction coefficient of $1.57 \times 10^4\,M^{-1}cm^{-1}$ (Sarin et al. 1981). The ninhydrin test is routinely used in SPPS.

After deprotection, a peptide resin sample is taken and tested in organic buffer with the ninhydrin reagent. In two to five minutes, an intense blue color develops on the beads and in the solution, unless the terminal amino acid is proline or benzyl aspartate, in which case there is a brown or reddish-brown color. After each coupling step, a sample is taken and heated for 5 to 10 minutes with the ninhydrin reagent. If any blue color develops on the beads or in the solution, an incomplete coupling is indicated. The rate and extent of the color development are very sensitive to heating time, temperature, and individual peptide. The blue color is not stable and decays within 10 to

15 minutes. Pyridine, which is one of the reagents used in this quantitative test, partially deprotects Fmoc amino acid resins. The deprotected amino acid resin may then react with the ninhydrin reagent, resulting in a higher Ruhemann's purple concentration than the true coupling efficiency warranted. Premature removal of the Fmoc group can be minimized by heating the reaction mixture for 5 minutes instead of 10 minutes and by adding two to three drops (0.02–0.04 mL) of glacial acetic acid to each resin sample.

In aqueous buffers, a blue chromophore is formed and is stable at 100°C for two hours, during which the quantitative analysis can easily be carried out. Because the curves do not show a stable plateau in organic medium, the rate and extent of color development are very sensitive to heating time, temperature, and individual peptide.

For example, ε'_{570} for HGly resin was $1.01 \times 10^4 M^{-1}, cm^{-1}$, and for HGlu(OBzl) resin, it was 1.25×10^4. The accuracy of the calculation of free amino groups by this method depends on the quality of the secondary method used to establish the extinction coefficient, such as amino acid analysis, picrate titration, and Edman end-group analysis. The precision of the assay on replicate samples was about ±4% over most of the analytical range, and the response curve was shown to be linear up to at least 1 mmol of amine per gram of resin.

For the routine monitoring of a synthesis, where an estimate of the extent of the coupling reaction is needed and where continued coupling to a constant endpoint is planned, the use of an average ε is usually adequate. Where an exact value for the percent of uncoupled amine is required, it becomes necessary to establish the extinction coefficient for the particular peptide resin under study. The rapid ninhydrin color test, as applied to samples of solid-phase resin peptides, makes it possible to detect incomplete couplings not discernible by amino acid analysis, to show diminishing amounts of free terminal amino groups by decreasing color intensities, and to obtain a positive test even though more than 99% of the terminal amino groups are reacted.

Picric acid (Diagram 5.1) was initially used for noninvasive monitoring of free amines (Gisin 1972). The resin is treated with picric acid to form a salt, rinsed to remove unbound acid, and then reacted with diisopropylethylamine (DIEA). The picric acid/DIEA chromophore is measured quantitatively at 362 nm (extinction coefficient 14,500 $M^{-1}cm^{-1}$) to determine the amine content in the resin (Sarin et al. 1981). This method is used in the amine concentration range of $1–20 \times 10^{-5} M$. Dilute acid labile groups are susceptible to removal by picric acid, making this technique incompatible with some Fmoc amino acid resins.

A method based on the condensation reaction of 2-hydroxy-1-naphthaldehyde with the free amino groups in the resin was reported (Esko et al. 1968). A quite stable aldimine (Schiff's base) absorbing at 420 nm is formed. The chromophore is then displaced from the polymer by an amine, e.g., benzylamine, and the amount of the soluble aldimine thus formed is determined. A 50-mg resin sample is required for an analysis.

Dorman (1969) described a simple, nondestructive procedure for quantitatively determining the completeness of solid-phase coupling reactions in conjunction with the use of N-t-butyloxycarbonyl (BOC) protected amino acids and active esters (Scheme 5.7). Following the coupling reaction and its subsequent washing steps, the resin peptide is treated with pyridine hydrochloride (Pyr.HCl) in dry methylene chloride to form the hydrochloride of any uncoupled resin and is then treated with triethylamine in DMF for three minutes to neutralize any hydrochloride formed. The filtrates are combined, chilled, acidified with excess dilute nitric acid, and titrated potentiometrically with standard silver nitrate. The equivalents of silver nitrate required represent the amount of uncoupled resin amine.

Dorman's test

Scheme 5.7

The success of this method rests on the ability of pyridine hydrochloride to form quantitatively the hydrochloride of uncoupled resin amine without causing premature removal of the t-butyloxycarbonyl protective group of the coupled amino acid. Unfortunately, Pyr.HCl/CH_2Cl_2 cleaves the o-nitrophenylsulfenyl (NPS) protective group at a substantial rate, thereby precluding the use of this reagent with an NPS protective group.

One method (Reddy and Voelker 1988) that essentially quantitates the unreactive N-terminal amino groups involves tritylation of the uncoupled amino groups with 4,4′-dimethoxytrityl chloride (DMTCl) in the presence of tetra-*n*-butylammonium perchlorate and subsequent detritylation by a mild acid (Scheme 5.8). Measurement of the color of the released trityl cation determines the coupling efficiency.

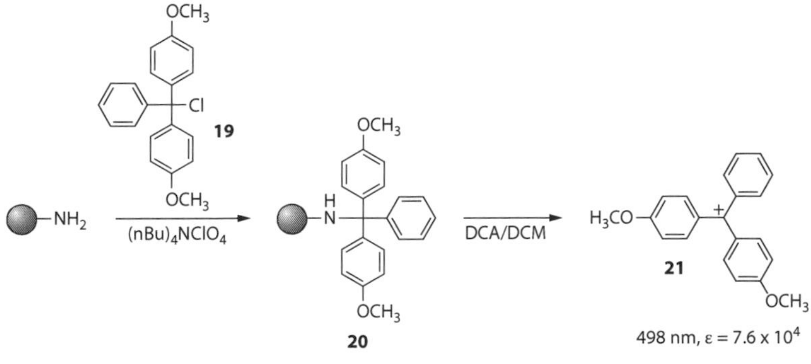

Scheme 5.8

The color of the released trityl cation measured spectrophotometrically at 498 nm ($\varepsilon = 7.6 \times 10^4$) determines the loading of amino functionality on the starting resin, which serves as a reference. Kinetic studies with support-bound unblocked isoleucine showed complete tritylation of amino groups in 30 seconds. The detritylation with 2% dichloroacetic acid (DCA) was complete within two minutes. The 2% DCA/CH_2Cl_2 did not release any BOC protecting group even after three hours of treatment. The solutions of DMTCl/CH_2Cl_2 and $(nBu)_4NClO_4$/CH_2Cl_2 were stable at room temperature for at least 45 days without any detectable loss of activity, as determined by comparative trityl assay with the freshly prepared reagents.

The advantages of this approach include the following:

1. The stable covalent attachment of the monitoring reagent allows extensive washings to remove the excess reagent, which makes the assay consistent and reliable.
2. The monitoring procedure is nondestructive and reversible in nature. Hence, the operations can be performed in the reactor itself, which makes the desired automation simpler.
3. Compared to the aliquot monitoring methods, sensitivity is increased because the total resin is employed for monitoring. Although the results were correlating closely with the Kaiser ninhydrin test, there were a few instances where this assay could detect unreacted amino groups of much less than 1%, where these could not be detected by the Kaiser test.

To quantitate the nonbasic or relatively hindered amines, nitro-phenylisothiocyanate-O-trityl (NPIT, Scheme 5.9) has been applied to measure the amount of a free amine group on resin based on the concentration of the released trityl cation concentration in solution (Chu and Reich 1995).

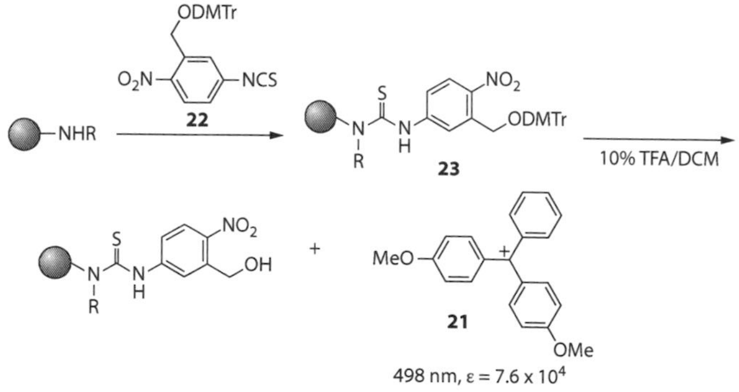

Scheme 5.9

A colorimetric method (Tyllianakis et al. 1994) for the direct determination of total solid-supported sulfhydryl, aldehydo, hydrazido, and N-hydroxysuccinimido carboxylate groups as well as of immobilized cysteine, tyrosine, and thyroxin using only the commercially available bicinchoninic acid/copper protein assay reagent was described (Scheme 5.10). The method is based on the ability of these groups to reduce Cu^{2+} to Cu^+, which forms a chelate complex with bicinchoninic acid absorbing at 562 nm. Each assay requires only one incubation step of the solids with the reagent for one hour at 60°C. The quantitation of the various groups is finally carried out through standard curves of appropriate substances.

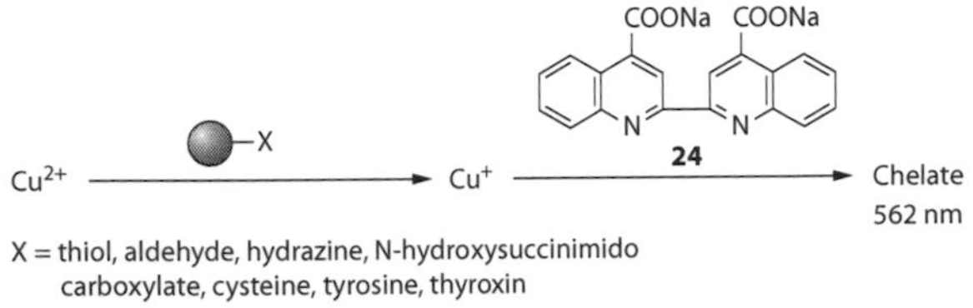

X = thiol, aldehyde, hydrazine, N-hydroxysuccinimido
carboxylate, cysteine, tyrosine, thyroxin

Scheme 5.10

The assays were found to be accurate, precise (inter assay CV less than 3%), and very sensitive, allowing the determination of nanomole quantities of functional groups per assay tube.

A simple colorimetric method (Ngo 1986) has been developed for the quantitative determination of reactive amino groups of a solid support. The method uses a reagent that reacts specifically and facilely with amino groups and concurrently introduces free sulfhydryl groups (Scheme 5.11). The latter can then be titrated colorimetrically by using the well-known Ellman's reagent. Thus, an excess of 2-iminothiolane (Traut's reagent) was reacted with amine-carrying solid supports. After removing unreacted 2-iminothiolane and keeping the resultant solid-phase sulfhydryl groups in a reduced state by using dithiothreitol, the solid supports, after being thoroughly washed, were then reacted with 5,5′-dithiobis-(2-nitrobenzoic acid) (DTNB) to quantify the sulfhydryl groups that were generated by reacting solid-phase amino groups with Traut's reagents. The DTNB reacts with sulfhydryl groups at pH 8 to produce 1 mol of 2-nitro-5-thiobenzoic acid per mole of sulfhydryl group. The 2-nitro-5-thiobenzoic acid is quantitated by UV spectroscopic measurement.

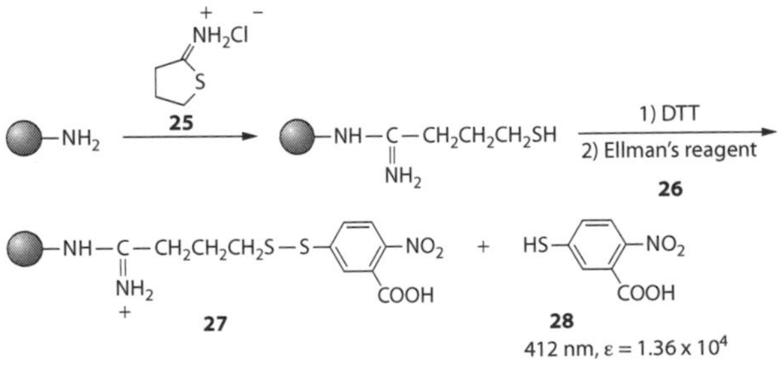

Scheme 5.11

Fmoc SPPS also offers a unique opportunity to noninvasively monitor N-α-amino group deprotection (Scheme 5.12). A fulvene-piperidine adduct is formed during piperidine deprotection. This adduct has UV absorption maxima at 267 nm ($\varepsilon = 17500$–18950 M^{-1} cm^{-1}), 290 nm ($\varepsilon = 5280$–5800 M^{-1} cm^{-1}), and 301 nm ($\varepsilon = 6200$–7800 M^{-1} cm^{-1}). These adsorption properties permit reasonably accurate determination of deprotection efficiency. Quantitative spectrophotometric monitoring of the deprotection of the Fmoc amino acid resins can also be used to determine the substitution level of amino acid per gram of resin (Heimer et al. 1981; Wade et al. 1991).

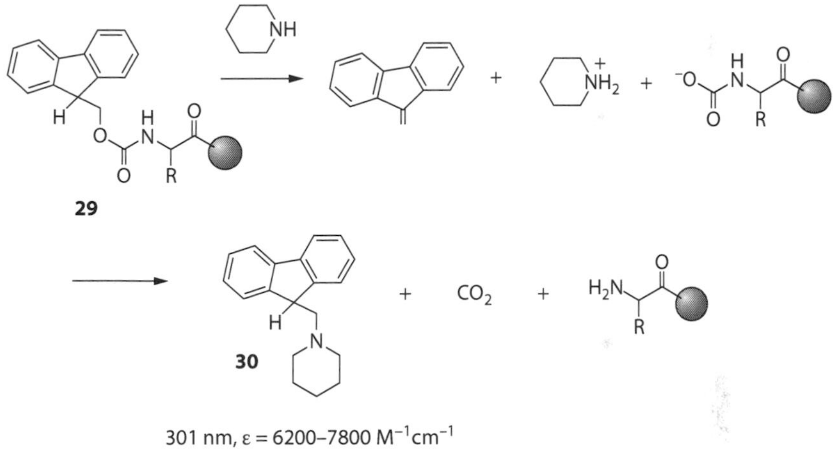

29

30

301 nm, $\varepsilon = 6200$–7800 $M^{-1}cm^{-1}$

Scheme 5.12

It has also been applied to the continuous-flow, solid-phase peptide synthesis (Meienhofer et al. 1979). The procedures of Fmoc synthesis and quantitative measurement have been analyzed in detail (Fontenot et al. 1991). An excellent review has summarized more specific information on Fmoc chemistry (Fields and Noble 1990).

The rate of attachment of Fmoc-containing Knorr linker to solid supports was quantitatively determined for various cross-linked polystyrene resins by Fmoc quantitation (Li et al. 1999). Results indicate that amide formation is favored on amino resins of smaller size and those with polyethyleneglycol (PEG), while both reaction rate and percentage of completion drop significantly when the reaction is performed in DMF in place of DCM.

Similar to the Fmoc-peptide chemistry, 4-nitrophenylethyloxycarbonyl (NPEOC) has been used as a protection group of amines in peptidylphosphonates synthesis (Campbell and Bermak 1994). The NPEOC group can be cleaved to form a stable nitrostyrene that has an intense absorbance at 308 nm with an extinction coefficient of 13,200 $M^{-1}cm^{-1}$.

5.4 Quantitative spectroscopic methods for organic functional groups

Methods previously developed for amine quantitation in SPPS can also be used in SPOS. However, SPOS involves many more organic functional groups. Quantitative analyses of these groups are needed in the reaction optimization stage.

5.4.1 Quantitation of aldehyde/ketone groups

On-resin quantitation of the loading in the starting resin and the yield of synthetic intermediate and product are required for reaction optimization in polymer-supported combinatorial and high-throughput parallel synthesis. A general principle for direct quantitation of the absolute amount of organic functional groups in resin-bound starting materials, intermediates, and products has been developed (Yan and Li 1997). The method employs fluorescence or UV-Vis spectroscopy and the availability of a large number of fluorescent dyes, which can be attached to various organic functional groups, to quantify organic reactions directly on resin. As the first application of this principle, a fluorescence method for rapid determination of the absolute amount of aldehyde or/and ketone groups on polystyrene and PS-PEG resins (Scheme 5.13) has been reported (Yan and Li 1997).

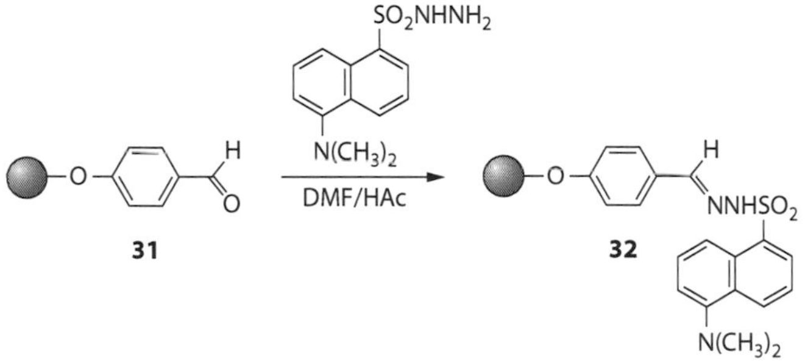

Scheme 5.13

To a suspension of formylpolystyrene or PS-PEG-CHO resin in DMF was added an approximately twofold excess of dansylhydrazine. The fluorescence spectra in Figure 5.1a and Figure 5.1b show the decrease of hydrazine concentration in the supernatant for reactions with formylpolystyrene and PS-PEG-CHO resins. After the reaction is completed, the consumed amount of dansylhydrazine is equal to the absolute amount of dye molecules covalently and noncovalently bound to the resin beads. The amount of hydrazine molecules noncovalently trapped in the resin matrix was corrected by subtracting the amount of dye molecules trapped in the

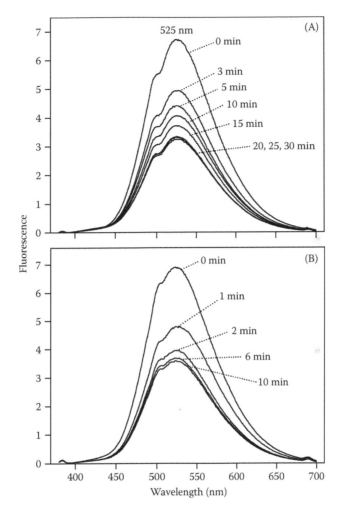

Figure 5.1 Fluorescence emission spectra of dansylhydrazine in the supernatant of the reaction mixture at various times for (A) formylpolystyrene resin and (B) TentaGel S-CHO resin. (Reproduced with permission from *J. Org. Chem.* 1997, 62, 9354. Copyright© 1997, American Chemical Society.)

same amount of 1% divinylbenzenepolystyrene or PS-PEG-Br resin under identical conditions.

To get the real-time reaction information during the time course of this reaction, single bead IR spectra were obtained at various times during the reaction (Figure 5.2). For formylpolystyrene resin (Figure 5.2A), the intensities of the C-H stretch band at 2728 cm⁻¹ and the carbonyl band at 1700 cm⁻¹ from the aldehyde decrease, whereas the intensities of the -N(CH₃)₂ stretch band at 2790 cm⁻¹ and the N-H band from the hydrazone at 3218 cm⁻¹ increase with time. These data unambiguously confirmed

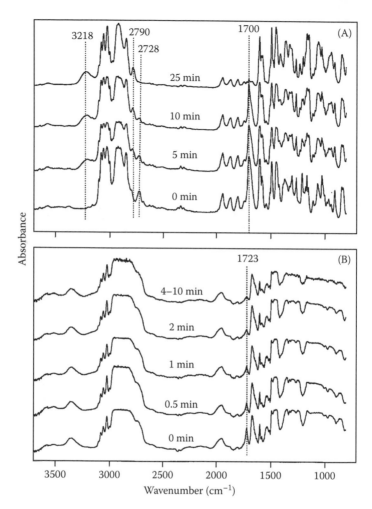

Figure 5.2 IR spectra of the single bead taken out of the reaction mixture at various times during the reaction for (A) formylpolystyrene resin and (B) TentaGel S-CHO resin. (Reproduced with permission from *J. Org. Chem.* 1997, 62, 9354. Copyright© 1997, American Chemical Society.)

that the decrease in hydrazine concentration in supernatant shown in Figure 5.1A is due to its reaction with the resin-bound aldehyde to form hydrazone on resin. Similarly, the reaction on PS-PEG resin was followed by the decrease of the intensity of the aldehyde carbonyl band at 1723 cm^{-1} (Figure 5.2B).

This method is simple to perform and accurately quantitates the absolute amount of aldehyde and aliphatic ketone groups on resin. It can be applied to 3 to 10 mg of resin sample (for loading 1 to 0.3 mmol/g) containing low micromolar amounts of functional groups.

The classical reagent 2,4-dinitrophenylhydrazine (DNPH) was used to colorimetrically detect aldehydes on several solid supports (Shannon and Barany 2004) based on ambient temperature reaction (Scheme 5.14). In the presence of dilute acid, DNPH reacts rapidly with resin-bound aldehydes, leading to the formation of a highly conjugated phenylhydrazone derivative, which is indicated by a red to dark-orange appearance at 25°C within one minute. In contrast, resins that are free of aldehydes or in which aldehyde functions have reacted completely retain their original color. The DNPH test was demonstrated for poly(ethylene glycol)-polystyrene (PEG-PS), aminomethyl polystyrene (AMP), cross-linked ethoxylate acrylate resin (CLEAR), and acryloylated O, O'-bis(2- aminopropyl)poly-(ethylene glycol) (PEGA) supports. The DNPH test allowed detection of aldehydes sensitively (18 μmol/g) with as little as 2 mg of aldehyde resin.

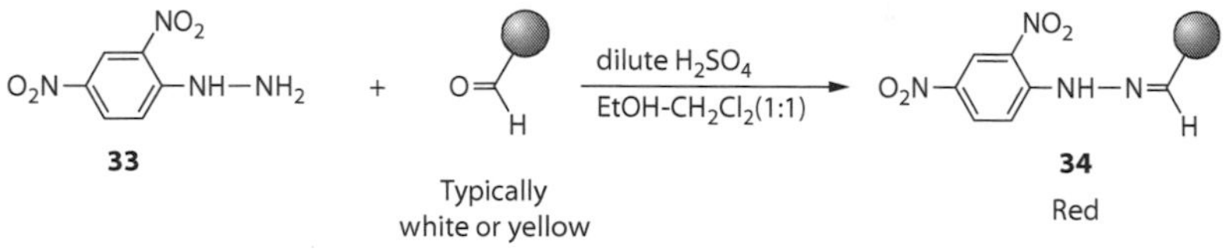

Scheme 5.14

A sensitive and specific color test was developed for the detection of the presence of resin-bound aldehyde groups using 4-amino-3-hydrazino-5-mercapto-1,2,4-triazole (Purpald). The reaction of gel-type resin-bound aldehydes with Purpald is shown in Scheme 5.15. Aldehyde resin turns dark brown to purple after a 5-minute reaction followed by a 10-minute air oxidation period. Resins that possess other functional groups (such as ketone, ester, amide, alcohol, and carboxylic acid) do not change color under the same conditions. The detection limit is 20 μmol/g for polystyrene-based aldehyde resins.

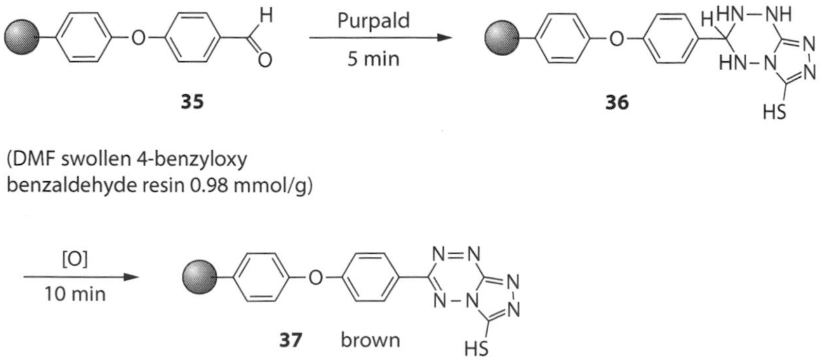

Scheme 5.15

5.4.2   Quantitation of hydroxyl groups on resin

Because the UV-Vis spectrometer is more stable, less expensive, and widely available, a UV spectroscopic method for rapid determination of the absolute amount of hydroxyl groups on polystyrene resin using 9-anthroylnitrile (Scheme 5.16) has been reported (Yan et al. 1999).

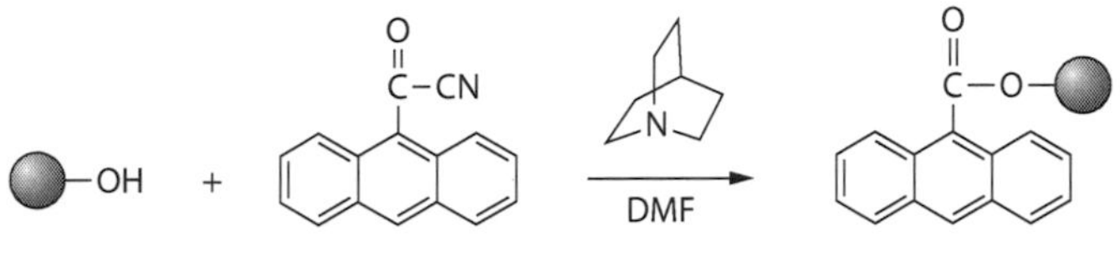

38

9-anthroylnitrile

39

Scheme 5.16

An approximately twofold excess of 9-anthroylnitrile in DMF was added to a suspension of Wang resin and to the same amount of plain polystyrene resin control (5–9 mg in 0.2 mL DMF with an excess amount of quinuclidine). The UV spectra of the supernatant for polystyrene control and the Wang resin were taken after 20 min. Figure 5.3a shows a decrease in dye concentration in the supernatant. The reaction occurs in the solid phase as shown by the single-bead FTIR spectra taken before and after 20 min. Figure 5.3b shows the disappearance of hydroxyl bands at 3458 and 3578 cm^{-1} and the formation of a new band at 1722 cm^{-1} after 20 min. The band at 1722 cm^{-1} is attributable to the ester carbonyl in product **39**. Single-bead IR spectra at various times during the reactions were also measured (Figure 5.4). Intensities of the O-H stretch bands at 3578 cm^{-1} and 3458 cm^{-1} decrease gradually, and the carbonyl band at 1722 cm^{-1} from the product increases with time. The time course again shows that the esterification is complete in 20 min.

The absolute amount of hydroxyl groups in various PS-based samples was then determined by the dye consumption deduced from UV spectroscopic measurements. The amount of dye molecules noncovalently trapped in the resin matrix was simultaneously corrected by subtracting the amount of dye molecules adsorbed in the same amount of 1% divinylbenzene polystyrene resin under identical conditions. The loading of hydroxyl was calculated based on the decreased UV-Vis absorbance. The results agreed with the loading values determined by the Fmoc-derivatization method. The advantages of this method are its simplicity, speed, and the requirement of low milligram amount of samples.

An immobilized fluorescent tracer molecule dansyl chloride (Scheme 5.17) is used to monitor the reaction on hydroxyl resins (Argopore Wang, PS Wang, and Argogel Wang) in parallel chemistry (Pina-Luis et al. 2004). Fluorescence measurements were obtained directly on solid phase (Figure 5.5). This method, with high sensitivity and a rapid procedure, has

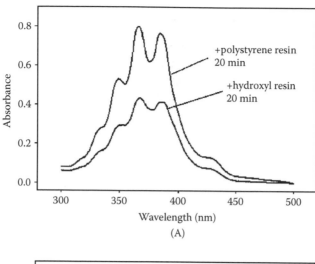

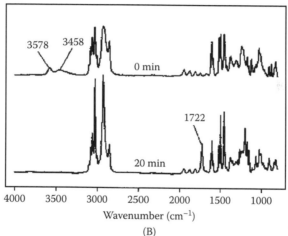

Figure 5.3 UV–visible absorption spectra of the reagent and single-bead FT-IR spectra. (A) UV–visible spectra of the supernatant that contains 9-anthroylnitrile-quinuclidine adduct mixed with polystyrene resin or Wang resin for 20 min. A 5-μL aliquot of supernatant was diluted into 2 mL of DMF for measurement. DMF was used as reference, and the path length was 1 cm. (B) Single-bead FT-IR spectra taken from the starting material Wang resin and the same resin after a 20-min reaction with 9-anthroylnitrile. The IR results show the formation of resin-bound ester carbonyl group and the disappearance of the hydroxyl group. All IR spectra were taken using the transmission mode at room temperature as described in the Experimental Section. (Reproduced with permission from *Anal. Chem.* 1999, 71, 4564. Copyright© 1999, American Chemical Society.)

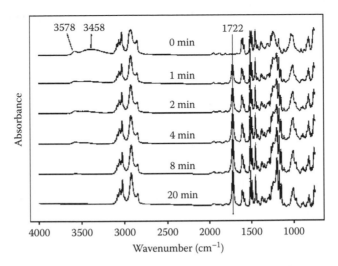

Figure 5.4 Single-bead FT-IR spectra at various times during the reaction shown in Scheme 16. Spectra were taken from a single flattened bead at 0, 1, 2, 4, 8, and 20 min after the initiation of the reaction. The hydrogen-bonded and unbonded hydroxyl stretching at 3458 and 3578 cm^{-1} disappears as the ester carbonyl band at 1722 cm^{-1} increases with time. (Reproduced with permission from *Anal. Chem.* 1999, 71, 4564. Copyright$^{©}$ 1999, American Chemical Society.)

been demonstrated to be a valuable tool for the quantitative determination of resin-bound hydroxyl groups, for studying reaction kinetics, and for continuously monitoring the progress of the conversion of the hydroxyl resins into chlorinated ones, as compared with the traditional methods in which the cleavage and subsequent analysis of the released products are necessary. The system can be extended to monitor a variety of reactions on solid supports, and in conjunction with a well-established technique such as flow analysis, basic studies on solid phase become possible.

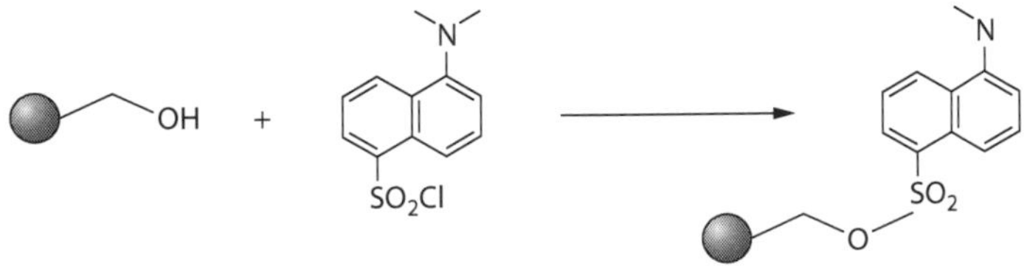

Scheme 5.17

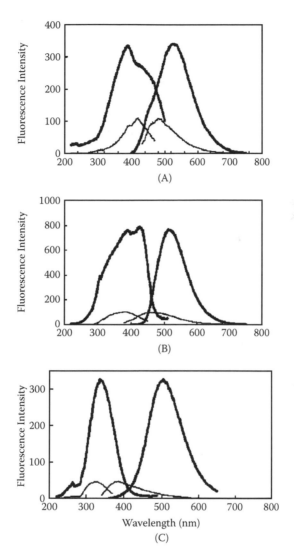

Figure 5.5 Fluorescence excitation and emission spectra of HO resins before (thin line) and after (thick line) reaction with dansyl chloride: (A) Argopore Wang, (B) PS Wang, and (C) Argogel Wang. (Reproduced with permission from *J. Comb. Chem.* 2004, 6, 391. Copyright© 2004, American Chemical Society.)

5.4.3   Quantitation of carboxyl groups on resin

PDAM reacts with carboxyl groups to form an ester product (Scheme 5.18). In the study on the solid-phase version of this reaction (Yan et al. 1999), the reaction was found to be fast and efficient, allowing the quantitative determination of solid-phase-bound carboxyl groups.

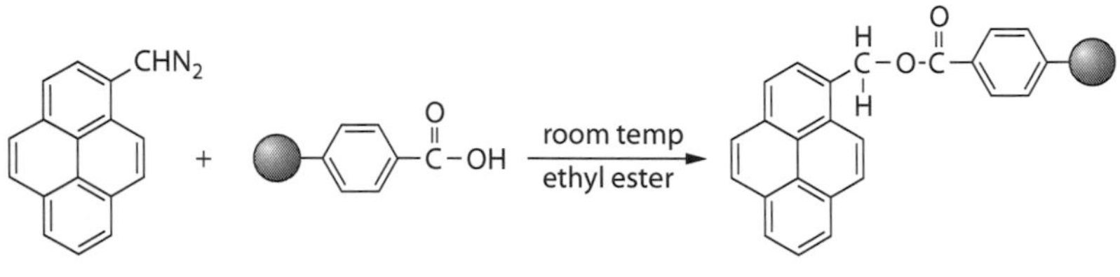

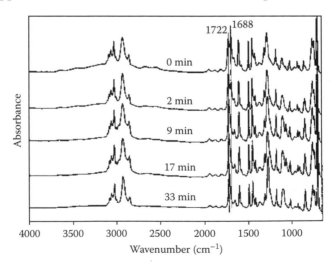

Scheme 5.18

The time course of this reaction was monitored by the single-bead FTIR method, as shown in Figure 5.6. The IR spectrum of the starting carboxyl resin (Figure 5.6, spectrum at 0 min) exhibits two carboxyl bands at 1688 and 1722 cm⁻¹ corresponding to associated and free carboxyl groups. After a reaction time of 33 min, the band at 1688 cm⁻¹ disappears while an ester carboxyl band in the product appears coincidentally also at 1722 cm⁻¹. Besides these spectral changes, the acid-OH group stretching signals at 2500–3500 cm⁻¹ also disappear after 33 min. The reaction was thus complete in 30–40 min.

Figure 5.6 Single-bead FT-IR spectra taken at various times during the reaction shown in Scheme 5.18. (Reproduced with permission from *Anal. Chem.* 1999, 71, 4564. Copyright© 1999, American Chemical Society.)

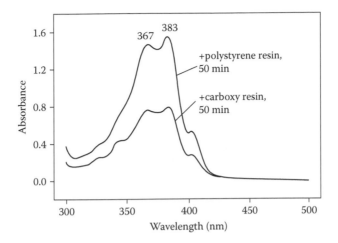

Figure 5.7 UV–visible absorption spectra of PDAM in the supernatant before and after a 50-min reaction. (Reproduced with permission from *Anal. Chem.* 1999, 71, 4564. Copyright© 1999, American Chemical Society.)

In order to ensure the completion of the reaction with carboxyl resins from various sources, all quantitation reactions were carried out for 50 min using an approximately twofold excess of PDMA. To calibrate noncovalently-bound PDMA molecules in the resin, plain polystyrene resin was used as a control. Figure 5.7 shows the UV-Vis absorption spectra of PDMA after mixing with carboxyl resin (lower spectrum) or polystyrene resin (top spectrum) for 50 min. A variety of resins (aminomethyl, prolinol-2-chlorotrityl, oxime, Rink acid, and Wang resins) were tested for their reactivity with PDMA, and no spectral change in the starting resins for formation of ester or amide products were detected by single-bead IR. Phenol resin has low reactivity toward PDMA.

In summary, a method using PDMA and UV-Vis spectroscopy for rapidly quantifying carboxyl groups in both PS and PS-PEG resins has been developed. Most organic functional groups except phenol do not interfere with the quantitation. The method provides a useful tool for assisting reaction optimization in combinatorial synthesis.

5.4.4   Quantitation of polymer-supported sulfhydryl groups

A sensitive and simple method is described for the quantitative determination of free sulfhydryl (SH) groups on polymer supports (Sharma et al. 1990). The method includes the reaction of 4,4′-dimethoxytrityloxy-S-(2-thio-5-nitro-pyridyl)-2-mercaptoethane (DTNPME) with polymer-supported sulfhydryl groups. After removal of excess reagent through washing, a polymer support is treated with perchloric acid to release the 4,4′-dimethoxytrityl cation from the polymer support into the solution.

The dimethoxytrityl cation (λ_{max} =498 nm, ε_{498} =70,000 $M^{-1}cm^{-1}$) is then quantified spectrophotometrically.

A method (Ellman 1959) has been reported for the qualitative and quantitative determination of sulfhydryl groups in simple (low molecular weight) sulfhydryl (SH) compounds and biological samples. Under certain conditions, this method can be used in the analysis of SPOS resin (e.g., on PS-PEG resins).

5.5   Combustion elemental analysis methods

Elemental analysis of resin-bound molecules has often been used for solid-phase synthesis papers (Frechet and Schuerch 1971; Mitchell et al. 1978; Frechet et al. 1979; N'Guyen and Boileau 1979; Forman and Sucholeiki 1995). The scope and value of this method is very clear. A systematic investigation on the role of combustion elemental analysis in the quantitative monitoring of solid-phase organic synthesis was reported (Yan et al. 1998). In this section, quantitative elemental analysis results obtained for eight resin-bound organic compounds are compared with those obtained from well-established spectrophotometric methods (Fmoc methods). In addition, a six-step peptoid synthesis and a dansylhydrazone synthesis were quantitatively monitored using elemental analysis. All results were corroborated by well-established spectrophotometric methods.

5.5.1   Analysis of resin-bound organic compounds

A series of resin-bound compounds were quantitatively analyzed by combustion elemental analysis methods (Diagram 5.3). In a carbon-hydrogen-nitrogen (CHN) analysis, resin compounds are combusted and oxidized completely to form CO_2, H_2O, and N_2, which are then quantitatively determined simultaneously on the basis of their different thermoconductivities relative to the carrier gas helium. All samples for CHN analysis were in reference to the standard acetanilide. 1% divinylbenzene (DVB)-polystyrene copolymer is the basic matrix for many resins used in SPOS. The theoretical C/H ratio of the plain 1% DVB-polystyrene resin based on the ratio of starting materials used for polymerization is 92.3/7.7. Elemental analysis of this resin in quadruple gave a ratio of 92.0±0.2 / 7.7±0.1, showing excellent reproducibility and accuracy. To further investigate the reproducibility of the method, a multiple analysis of compounds **46–49** shown in Diagram 5.3 was carried out for their nitrogen contents (Table 5.1). The reproducibility of the data is very good, as shown by the very small variations. It is important to point out that only 2 mg of resin is required for a CHN analysis and that each analysis takes less than 20 min in an automated fashion. In routine analysis, a single analysis or a duplicate measurement should be sufficient.

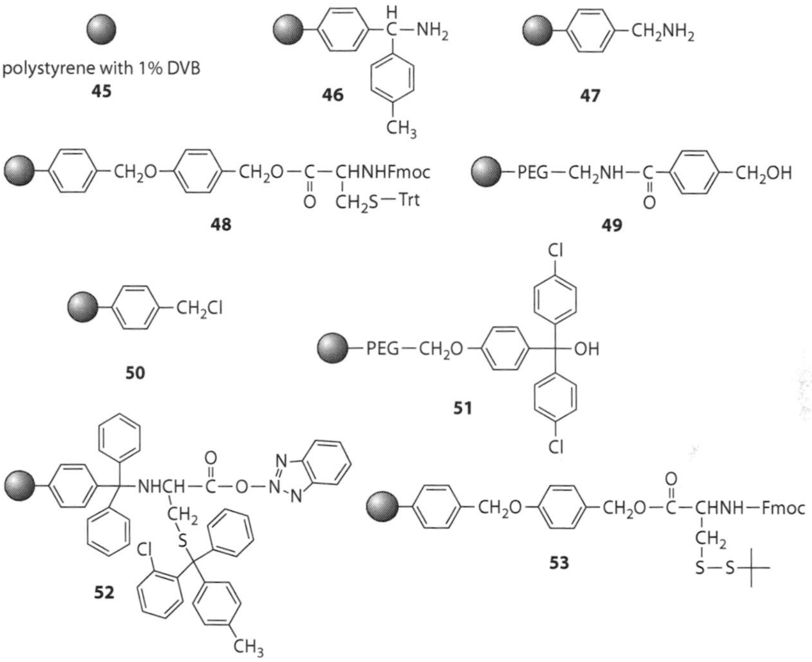

Diagram 5.3

In addition to nitrogen quantitation, a series of resin samples containing sulfur and chlorine (**50–53**) were also analyzed (Table 5.1). All the expected values, including those for the nitrogen quantitation mentioned above, came from well-established spectrophotometric analysis, such as analysis of the cleaved fluorenylmethoxy-carbonyl (Fmoc) and N-hydroxylbenzotriazole (HOBt) groups, except that the value for plain polystyrene is based on the ratio of starting materials for polymerization. Standard deviations from the expected value for most samples are below 5%, a level acceptable for quantitative analysis in SPOS.

5.5.2   Quantitatively monitoring SPOS reactions

To apply the elemental analysis method in the quantitatively monitoring of multistep SPOS, the synthesis of a dansylhydrazone from a resin-bound aldehyde was investigated (Scheme 5.13). The nitrogen content was increased from <0.2 to 22.7 mg/g after the reaction (Table 5.2), showing a 93.4% conversion, which is also confirmed by single-bead IR and a dye-coupling quantitation method.

The successful monitoring of a series of peptoid synthesis reactions (Scheme 5.19) further substantiated the potential of the method for a quantitative monitoring of solid-phase organic synthesis. The starting material Rink

Table 5.1 Elemental Analysis of Resin-Bound Compounds: Reproducibility and Accuracy

Compound	Weight (mg)	Element	# of Expt.	Value (mg/g)	Expected[a] (mg/g)
45	2	C/H	4	92.0±2/77±1	92.3/77
46	2	N	4	6.7±0.2	6.44
47	2	N	4	14.4±0.3	14.7
48	2	N	4	9.9±0.2	9.1
49	2	N	4	3.8±0.2	3.9
50a[b]	15–20	Cl	2	36.5±0.01	36.1
50b	15–20	Cl	2	25.7±0.01	26.6
51	15–20	Cl	2	13.6±0.1	13.3
52	15–20	Cl	2	22.9±0.2	22.1
52	20–35	S	2	22.5±0.3	20.2
53	20–35	S	2	34.8±0.4	32.0

Note: [a] Theoretical value for 45 and spectrophotometric measurement after derivatization.
  [b] 50a and 50b are from two different batches.

Table 5.2 Elemental Analysis of Resin-Bound Compounds: Reaction Monitoring

Compound	Weight (mg per expt)	Element	Value (vs. theoretical) (mg/g resin)	Yield (N or Br analysis)	Yield NPIT
31	2	N	<0.2	–	–
32	2	N	22.7 (24.3)	93.4% [o]	–
54	25–35	Br	–	–	–
55	25–35	Br	0 (0.0)	–	109% [o]
56	25–35	Br	37.1 (36.7)	101% [o]	negative [s]
57	25–35	Br	0.12 (0.0)	97% [s]	100.5% [o]
58	25–35	Br	32.3 (34.9)	92.5% [o]	negative [s]
59	25–35	Br	0.08 (0.0)	complete [s]	78% [o]
60	25–35	Br	24.3 (32.3)	75.2% [o]	negative [s]
61 + 62	25–35	Br	0.02 (0.0)	complete [s]	30% [o]

Note: o–overall yield; s–one-step yield.

amide AM resin (**54**, 0.44 mmol/g) was first activated and quantitated by the cleavage of the Fmoc chromophore. The loading of the resulting amine resin is increased to 0.48 mmol/g as a result of the altered resin/functional group ratio after the leaving of Fmoc. In all SPOS, it is necessary to re-calculate the loading after each synthetic step because of the changed resin/functional group ratio. The increased weight from the new group was added to the base weight, and the original loading in terms of mmol/g was re-calculated after each reaction step. The six-step synthetic reaction was quantitatively monitored by both the bromine quantitative analysis and the nitrophenylisothiocyanate-O-trityl (NPIT) quantitation (Chu and Reich 1995). The disappearance of the primary and the secondary amines was quantitatively correlated with the formation of the brominated compounds as determined by both the NPIT test and the bromine quantitation (Table 5.2).

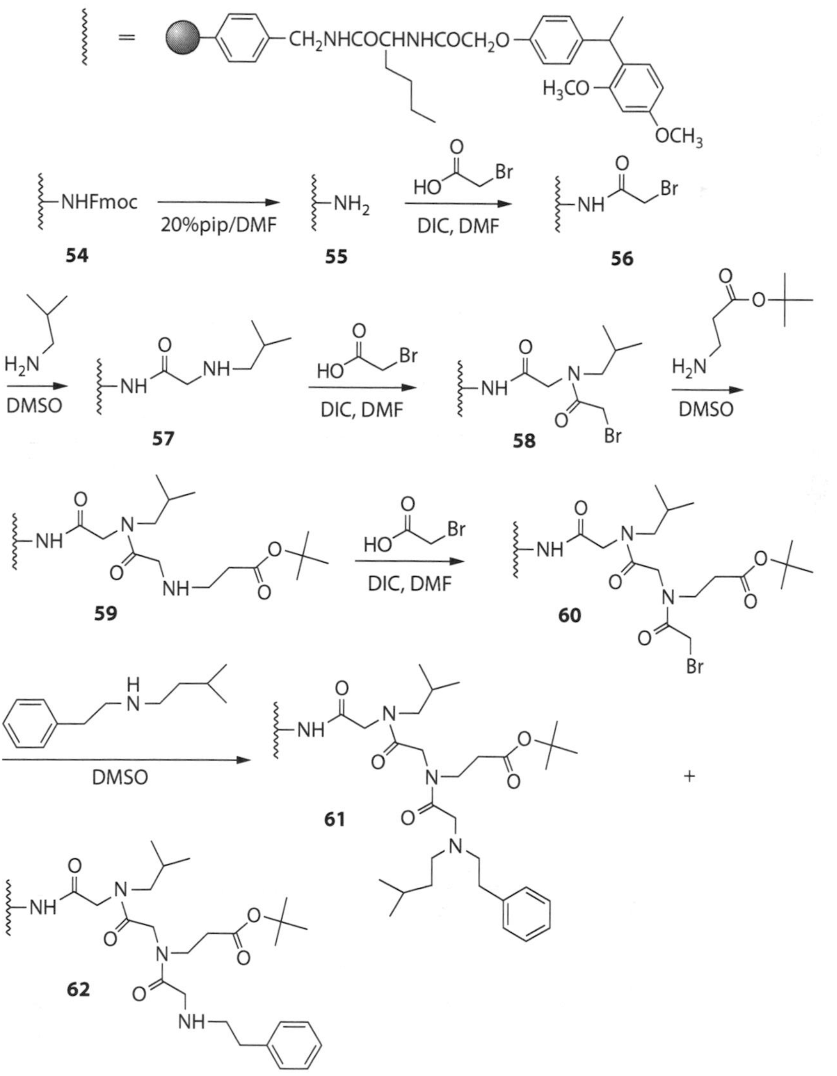

Scheme 5.19

In the last step, the bromine test suggests a quantitative conversion of **60** to **62**. However, the NPIT test suggests the presence of an only partial amine product. In order to investigate the inconsistency, end products were cleaved from resin and analyzed by MS and HPLC. MS confirmed the presence of both **61** and **62**, and HPLC indicated the ratio of **61/62** to be 1/1.82. The cause for the formation of side product **61** is the presence of a 6% impurity (phenylethamine) in the reagent in the last step of synthesis. This result indicates that the purity of reagents is crucial when the impurity is more reactive. This type of on-resin analysis effectively helps facilitate the reaction optimization steps required in developing on-resin synthetic processes.

Based on these data, the elemental analysis method is, in general, reproducible and accurate in the quantitative analysis of resin-bound organic compounds. Considering the severe shortage of on-resin quantitation methods at present, it adds a valuable tool for monitoring solid-phase organic synthesis steps that cannot be quantitated by current methods.

It should be emphasized that it is important to wash and dry the sample completely before the elemental analysis, because the resin-trapped reagent or solvent molecules will cause variable results. In practical considerations, although the combustion elemental analysis method is a destructive method and is not always sample efficient, it will prove to be of certain utility in SPOS.

5.6   *Electrochemical methods*

A noninvasive method for the monitoring and quantitative determination of basic functionality (pK_{HB^+} >7) on solid support has been described (Patek et al. 1998). This method is based on complete protonation of basic functionality after treatment of the resin with a large excess of 1% $HClO_4$, which provides an easily detectable perchlorate anion. After thorough washing, the bound anion is eluted with a suitable base and quantified potentiometrically with a perchlorate ion-selective electrode (ISE). The amount of basic component is calculated from an output potential that is logarithmically proportional to the activity of the selected ion in the sample phase via the Nernst equation.

Using the mV-meter with a resolution of 0.1 mV, one should be able to read changes in concentration of 0.4%. However, for practical purposes, a resolution of 1 mV will suffice. Because of the dependence of electrochemical potential on temperature, samples and standards must be at the same temperature at the instant of the measurement. A 1°C difference in temperature will result in a 3% error at the 10^{-4} M level.

It was noticed that there are some limitations in quantitative determination of basic functionality. One obvious problem inherent in this method is a salt hydrolysis effect. Because the procedure for perchlorate determination involves repetitive washings with distilled water, solvolysis of ammonium salt will occur, which is dependent upon the dissociation constant of the protonated base and the autoprotolytic constant (pK_{Ap}) of the solvent used. In general, as can be expected from the corresponding pK_{HB^+}'s, most primary, secondary, and tertiary amines, amidines, guanidines, imidazoles, and electron-rich dialkylaminobenzenes behave normally in aqueous solution, whereas pyridines and anilines are not suitable basic functionalities for quantitative determination. Notably, indol-type compounds are completely "invisible" with such a technique in an aqueous solution. In addition, attempts to extend the method to the aminomethylphenyl and p-methylbenzhydrylamine (MBHA) types of resins were not successful. Unfavorable swelling properties of these resins in aqueous media and a low tolerance of the ISE to organic solvents are the main reasons for the failure of this method.

In conclusion, ISE methodology can be used to quantify a variety of basic functionalities and their relative changes during a reaction on solid support. Limitations of this methodology include the use of water-compatible solid supports and basic functionality possessing a pK_{HB^+} within a certain range. A significant advantage of the described method is that it is simple and noninvasive compared to traditional "cleave-and-analyze" techniques. The concept of using ISE to monitor the progress of reactions on solid phase offers another attractive tool to solid-phase organic and peptide chemists.

Conventionally, the loading of Merrifield resins is determined by the elemental analysis of incorporated chlorine or titration of pyridinium chloride after hot pyridine treatment of the Merrifield resin. Because most of the mass of the MicroTube™ reactors or Chiron crowns is polypropylene and the percentage of incorporated chlorine in Merrifield MicroTube™ reactors is extremely small, elemental analysis would not be accurate. Titration is more tedious and requires more Merrifield MicroTube™ reactors for analysis. A potentiometric method that permits a measurement of chloromethylation using chloride ion selective electrode (ISE) was developed (Li et al. 1998). The method involves the formation of pyridinium chloride after hot pyridine treatment of the Merrifield MicroTube™ followed by the quantitation of chloride ions by chloride ion selective electrode. This method is simple, convenient, and very accurate. It gives accurate analytical results in a linear range of 5 to 150 nmol chloride per Merrifield MicroTube™ reactor. Typical Merrifield MicroTube™ reactors have a loading of 30 µmol ± 10% RSD (Relative Standard Deviation).

Another method for monitoring amine in solid-phase peptide synthesis was described (Brunfeldt et al. 1969, 1972). The method is based upon a potentiometric titration of the whole batch of resin-bound peptide suspended in a mixture of methylene chloride and glacial acetic acid 1/1 (v/v). Perchloric acid dissolved in glacial acetic acid was used as titrant.

The analysis was done in an automated fashion by connecting to the synthesizer.

Classic organic analysis methods can be used when the resin-bound compounds can be converted into the proper form in solution. One example is the Volhard titration analysis of chlorine content in Merrifield resins (Lu et al. 1981).

5.7   On-resin X-ray, EPR, SERS, and fluorescence methods

SPOS reactions were monitored by energy-dispersive X-ray spectroscopy (EDS) in the environmental scanning electron microscope (ESEM) (Carlton et al. 1997). EDS in the ESEM has the advantage of minimal sample size, speed, and simplicity because the analyses are performed without special specimen preparation. The progress of a two-step synthetic transformation was followed using EDS-ESEM by the presence of sulfur and bromine peaks in the products of each synthetic step. In another study, the elemental analysis of single-bead-bound library member by EDS was shown to be a reliable and fast method of analysis (Neilly and Hochlowski 1999). X-ray photoelectron spectroscopy (XPS) has also been used to monitor reactions on solid phase (Yoo et al. 1999). These X-ray-based methods require the involvement of heteroatom-containing molecules.

Polymer-bound peptide aggregation and its dependence on solvent and resin loading were studied by EPR (Cilli et al. 1997). Model low and highly peptide-loaded resins containing an aggregating sequence were labeled with a paramagnetic amino acid derivative and were studied regarding their solvation behavior in different solvent systems. EPR spectroscopy has allowed the detection of differentiated levels of peptide chain aggregation as a function of solvent and resin loading.

Fluorescent beads were studied by using bead suspension in a conventional cuvette method (Scott and Balasubramanian 1997). The range of linearity of on-bead fluorescence is shown to be dependent on both the extent of loading and the Stokes shift of the fluorophore. Fluorescence quenching was found in this study. However, suspension measurement allows heavy interbead quenching. To avoid such complications, fluorescence was recorded from a single bead (Yan et al. 1999). This study unambiguously demonstrated the intrabead fluorescence self-quenching when the Stokes shift of the fluorophore is small (when the absorption and emission spectra overlap).

Surface-enhanced Raman scattering (SERS) used for direct, label-free, and surface-selective detection of compounds bound to a single polystyrene bead in a few seconds has been reported for the first time (Schmuck et al. 2007). This method of on-bead detection using SERS is about 10^6–10^7

times more sensitive than a conventional Raman spectrum recorded in solution, allowing even 50 femtomoles of an artificial peptide receptor CBS-Lys-Lys-Phe-NHR attached to the TentaGel bead within a few seconds. SERS microspectroscopic mapping, that is, the combination of Raman microscopy with a spatially resolved detection, should allow the screening of entire combinatorial libraries.

5.8   Summary

The lack of quantitation methods is the major weakness of SPOS. The optimization of combinatorial chemistry, however, requires the development of more-effective on-support quantitative methods. These methods should have the following features:

1. They target a diverse set of organic functional groups.
2. They have high sensitivity and, therefore, require as small amount of a sample as possible.
3. They are fast and convenient for ordinary laboratories.

Although most organic functional groups can react with a reagent containing an Fmoc group to form an Fmoc derivative, this process takes a long time, requires more samples and two-step reactions, and is not applicable for normal SPOS scale. The dye consumption method has proved to be a useful method for quantitation.

Many classic analytical methods, such as electrochemical methods, EDS, EPR and elemental analysis, can also be used to perform quantitative analysis.

References

Brunfeldt, K., Christensen, T., Villemoes, P. 1972. Automatic monitoring of solid phase synthesis of a decapeptide. *FEBS Lett.* 22, 238.

Brunfeldt, K., Roepstorff, P., Thomsen, J. 1969. Process control in the solid phase peptide synthesis by titration of frmm amino groups. *Acta Chem. Scand.* 23, 2906.

Campbell, D. A., Bermak, J. C. 1994. Solid-phase synthesis of peptidylphosphonates: Rapid fluorescence determination of the absolute amount of aldehyde and ketone groups on resin supports. *J. Am. Chem. Soc.* 116, 6039.

Carlton, R. A., Orton, E., Lyman, C. E., Roberts, J. E. 1997. Qualitative analysis of solid phase synthesis reaction products by X-ray spectrometry. *Microsc. Microanal.* 3, 520.

Cho, J. K., White, P. D., Klute, W., Dean, T. W., Bradley, M. 2003. Self-indicating resins: Sensor beads and in situ reaction monitoring. *J. Comb. Chem.* 5, 632.

Christensen, T. 1979. A qualitative test for monitoring coupling completeness in solid phase peptide synthesis using chloranil. *Acta Chem. Scand.* B 33, 763.

Chu, S. S,.Reich, S. H. 1995. NPIT: A new reagent for quantitatively monitoring reactions of amines in combinatorial synthesis. *Bioorg. & Med. Chem. Lett.* 5, 1053.

Cilli, E. M., Marchetto, R., Schreier, S., Nakaie, C. R. 1997. Use of spin label EPR spectra to monitor peptide chain aggregation inside resin beads. *Tetra. Lett.* 38, 517.

Claerhout, S., Ermolat'ev, D. S., Van der Eycken, E. V. 2008. A new colorimetric test for solid-phase amines and thiols. *J. Comb. Chem.* 10, 580.

Cournoyer, J. J., Kshirsagar, T., Fantauzzi, P. P., Figliozzi, G. M., Makdessian, T., Yan, B. 2002. Color test for the detection of resin-bound aldehyde in solid-phase combinatorial synthesis. *J. Comb. Chem.* 4, 120.

Dorman, L. C. 1969. A non-destructive method for the determination of coupling reactions in solid phase peptide synthesis. *Tetra. Lett.* 10, 2319.

Ellman, G. L. 1959. Tissue sulfhydryl groups. *Arch. Biochem. Biophys.* 82, 70.

Esko, K., Karlsson, S., Porath, J. 1968. A method for determining free amino groups in polymers with particular reference to Merrifield synthesis. *Acta Chem. Scand.* 22, 3343.

Felix, A. M., Jimenez, M. H. 1973. Rapid fluorometric detection for completeness in solid phase coupling reactions. *Anal. Biochem.* 52, 377.

Fields, G. B. and Noble, R. L. 1990. Solid phase peptide synthesis utilizing 9-fluorenylmethoxycarbonyl amino acids. *Int. J. Peptide Proteins Res.* 35, 161.

Fontenot, J. D., Ball, J. M., Miller, M. A., David, C. M., Montelaro, R. C. 1991. A survey of potential problems and quality control in peptide synthesis by the fluorenylmethoxycarbonyl procedure. *Peptide Res.* 4, 19.

Forman, F. W., Sucholeiki, I. 1995. Solid-phase synthesis of biaryls via the Stille reaction. *J. Org. Chem.* 60, 523.

Frechet, J. M. J., de Smet, M. D., Farrall, M. J. 1979. Application of phase-transfer catalysis to the chemical modification of cross-linked polystyrene resins. *J. Org. Chem.* 44, 1774.

Frechet, J. M., Schuerch, C. 1971. Solid-phase synthesis of oligosaccharides. I. Preparation of the solid support: Poly[p-(1-propen-3-ol-1-yl)styrene]. *J. Am. Chem. Soc.* 93, 492.

Gaggini, F., Porcheddu, A., Reginato, G., Rodriquez, M., Taddei, M. 2004. Colorimetric tools for solid-phase organic synthesis. *J. Comb. Chem.* 6, 805.

Garden, J. II., Tometsko, A. M. 1972. Fluorometric method for quantitative determination of free amine groups in peptide-containing Merrifield resins. *Anal. Biochem.* 46, 216.

Gisin, B. F. 1972. The monitoring of reactions in solid-phase peptide synthesis with picric acid. *Anal. Chim. Acta.* 58, 248.

Hancock, W. S., Battersby, J. E. 1976. A new micro-test for the detection of incomplete coupling reactions in solid-phase peptide synthesis using 2,4,6-trinitrobenzene-sulphonic acid. *Anal. Biochem.* 71, 260.

Heimer, E. P., Chang, C., Lambros, T., Meienhofer, J. 1981. Stable isolated symmetrical anhydrides of Nα-9-fluorenylmethyloxycarbonylamino acids in solid-phase peptide synthesis. *Int. J. Peptide Proteins Res.* 18, 237.

Kaiser, E., bossinger, C. D., Colescott, R. L., Olsen, D. B. 1980. Color test for terminal prolyl residues in the solid-phase synthesis of peptides. *Anal. Chim. Acta* 118, 149.

Kaiser, E., Colescott, R. L., Bossinger, C. D. and Cook, P. I. 1970. Color test for detection of free terminal amino groups in the solid-phase synthesis of peptides. *Anal. Biochem.* 34, 595.

Kay, C., Lorthioir, O. E., Parr, N. J., Congreve, M., McKeown, S. C., Scicinski, J. J., Ley, S. V. 2001/2002. Solid-phase reaction monitoring: Chemical derivatization and off-bead analysis. *Biotech. & Bioeng. (Comb. Chem.)* 71, 110.

Krchnak, V., Vagner, J., Lebl, M. 1988b. Noninvasive continuous monitoring of solid-phase peptide synthesis by acid-base indicator. *Int. J. Peptide Proteins Res.* 32, 415.

Krchnak, V., Vagner, J., Safar, P., Lebl, M. 1988a. Noninvasive continuous monitoring of solid-phase peptide synthesis by acid-base indicator. *Collect. Czech. Chem. Commun.* 53, 2542.

Li, R., Xiao, X., Czarink, A. W. 1998. Merrifield microtube™ reactors for solid phase synthesis. *Tetra. Lett.* 39, 8581.

Li, W., Xiao, X., Czarnik, A. W. 1999. Kinetic comparison of amide formation on various crosslinked polystyrene resins, *J. Comb. Chem.* 2, 151.

Lu, G. S., Mojsov, S., Tam, J. P., Merrifield, R. B. 1981. Improved synthesis of 4-alkoxybenzyl alcohol resin. *J. Am. Chem. Soc.* 46, 3433.

Mařík, J., Aimin Song, A., Kit S. Lam, K. S. 2003. Detection of primary aromatic amines on solid phase. *Tetra. Lett.* 44, 4319.

Meienhofer, J., Waki, M., Heimer, E. P., Lambros, T. J., Makofske, R. C., Chang, C. 1979. Solid phase synthesis without repetitive acidolysis, preparation of leucyl-alanyl-glycyl-valine using 9-fluorenylmethyloxycarbonylamino acids. *Int. J. Peptide Proteins Res.* 13, 35.

Mitchell, A. R., Kent, S. B. H., Engelhard, M. Merrifield, R. B. 1978. A new synthetic route to tert- butyloxycarbonylaminoacyl-4-(oxymethyl)phenylacetamidomethyl-resin, an improved support for solid-phase peptide synthesis. *J. Org. Chem.* 43, 2845.

Neilly, J. P., Hochlowski, J. E. 1999. Elemental analysis of individual combinatorial chemistry library members by energy-dispersive X-ray spectroscopy. *Appl. Spectr.* 53, 74.

Ngo, T. T. 1986. Colorimetric determination of reactive amino groups of a solid support using Traut's and Ellman's reagents. *Appl. Biochem. Biotech.* 13, 213.

N'Guyen, T. D., Boileau, S. 1979. Phase transfer catalysis in the chemical modification of polymers. Part II. *Tetra. Lett.* 28, 2651.

Patek, M., Bildstein, S., Flegelova, Z. 1998. Monitoring solid phase reactions with ion-selective electrode. *Tetra. Lett.* 39, 753.

Pina-Luis, G., Badía, R., Díaz-García, M. E. Rivero, I. A. 2004. Fluorometric monitoring of organic reactions on solid phase. *J. Comb. Chem.* 6, 391.

Reddy, M. P., Voelker, P. J. 1988. Novel method for monitoring the coupling in solid phase peptide synthesis. *Int. J. Peptide Protein Res.* 31, 345.

Sarin, V. K., Kent, S. B. H., Tam, J. P., Merrifield, R. B. 1981. Quantitative monitoring of solid-phase peptide synthesis by the ninhydrin reaction. *Anal. Biochem.* 117, 147.

Schmuck, C., Wich, P., Kuestner, B., Kiefer, W., Schluecker, S. 2007. Direct and label-free detection of solid-phase-bound compounds by using surface-enhanced Raman scattering microspectroscopy. *Angew. Chem. Int. Ed.* 46, 4786.

Scott, R. H., Balasubramanian, S. 1997. Properties of fluorophores on solid phase resins: Implications for screening, encoding and reaction monitoring. *Bioorg. Med. Chem. Lett.* 7, 1567.

Shah, A., Pahman, S. S., de Biasi, V., Camiller, P. 1997. Development of colorimetric method for the detection of amines bound to solid support. *Anal. Commun.* 34, 325.

Shannon, S.K., Barany, G. 2004. Colorimetric monitoring of solid-phase aldehydes using 2,4-dinitrophenylhydrazine *J. Comb. Chem.* 6, 165.

Sharma, P., Sathyanarayana, S., Kumar, P., Gupta, K. C. 1990. A spectrophotometric method for the estimation of polymer-supported sulfhydryl groups. *Anal. Biochem.* 189, 173.

Troll, W., Canan, R. K. 1953. A modified photometric ninhydrin method for the analysis of amino and imino acids. *J. Biol. Chem.* 200, 803.

Tyllianakis, P. E., Kakabakos, S. E., Evangelatos, G. P., Ithakissios, D. S. 1994. Direct colorimetric determination of solid-supported functional groups and ligands using bicinchoninic acid. *Anal. Biochem.* 219, 335.

Vojkovsky, T. 1995. Detection of secondary amines on solid phase. *Peptide Res.* 8, 236.

Wade, J. D., Bedford, J., Sheppard, R. C., Tregear, G. W. 1991. DBU as an N^α-deprotecting reagent for thefluorenyl-methoxycarbonyl group in continuous flow solid-phase peptide synthesis. *Peptide Res.* 4, 194.

Yan, B., Jewell, Jr., C. F., Myers, S. W. 1998. Quantitatively monitoring of solid-phase organic synthesis by combustion elemental analysis. *Tetrahedron* 54, 11755.

Yan, B, Li, W. 1997. Rapid fluorescence determination of the absolute amount of aldehyde and ketone groups on resin supports. *J. Org. Chem.* 62, 9354.

Yan, B, Liu, L., Astor, C. A., Tang, Q. 1999. Determination of the absolute amount of resin-bound hydroxyl or carboxyl groups for the optimization of solid-phase combinatorial and parallel organic synthesis. *Anal. Chem.* 71, 4564.

Yan, B., Martin, P., Lee, J. 1999. Single-bead fluorescence microspectroscopy: Detection of self-quenching in fluorescence-labeled resin beads. *J. Comb. Chem.* 1, 78.

Yoo, S.-E., Gong, Y.-D., Seo, J.-S., Sung, M.-M., Lee, S. S., Kim, Y. 1999. X-ray photoelectron spectroscopy analysis of solid-phase reactions using 3-brominated Wang resin. *J. Comb. Chem.* 1, 177.

Zhang, B., Li, X.F., Yan, B. 2008. Advances in HPLC detection-towards universal detection. *Anal. Bioanal. Chem.* 390, 299.

chapter six

Quality control of combinatorial libraries

6.1 Introduction

The value of biological screening data is highly dependent on the quality of the compound collections (Letot et al. 2005). The quality of combinatorial libraries and ways to determine and improve the purity and yield of desired compounds are of utmost importance (Yan et al. 2003). The demand for high-throughput analytical characterization of combinatorial libraries has prompted the development of more rapid methods to keep pace with compound production. Recent progress has focused on the development of parallel separation methods, multiplexed detector interfaces, fast LC (such as UPLC), and synergistic combinations of various detectors possessing complementary selectivity. Detection methods investigated for the high-throughput characterization of small-molecule combinatorial libraries are summarized in Table 6.1 (Fitch 1999; Kenseth and Coldiron 2004).

The characterization of all chemical library members may seem unnecessary, because it may be argued that an examination of the active compounds is necessary. Although it may be economical to do so, the lack of hits may mean that the tested compounds are not active or that the synthesis had failed. Such operations could mislead researchers into overestimating the diversity of molecules they explore. Therefore, it is important to confirm the desired diversity present in a library. The application of MS in such a confirmation study has recently been reviewed (Kassel 2001; Triolo et al. 2001; Shin and Breemen 2001; Cheng and Hochlowski 2002).

Although biological assays can be carried out on beads directly, in-solution assays are more practical. If the in-solution assay is the approach, the characterization of library members should be performed after cleavage from solid supports. The compound libraries to be analyzed are either a discrete compound library produced via parallel synthesis or a one-bead-one-compound library made through combinatorial synthesis such as split-and-pool synthesis (Furka et al. 1988, 1991; Lam et al. 1991; Houghten et al. 1991). In split-and-pool synthesis, discrete compounds are synthesized at the single-bead level. If the pooled libraries are examined as a whole, the analysis of mixtures is encountered. Various approaches

Table 6.1 Detection Methods Investigated for the High-Throughput Characterization of Small-Molecule Combinatorial Libraries

Detection Method	Strengths	Limitations
UV absorbance	Good sensitivity and linearity of response	Provides only relative purity measurements
	Easily integrated online with separation systems	Compounds must contain UV chromophore
		Response factors depend on extinction coefficients of compounds
Evaporative	Response factors of diverse compounds less structurally dependent than UV detection	Provides only relative purity measurements
light scattering	Easily integrated online with separation systems	Diminished response for compounds with MW < 300
Chemiluminescent	Universal response to number of moles of nitrogen structure	Compounds must contain nitrogen to be detected
nitrogen	Can provide quantitative purity measurements when used with internal or external standards	Certain rare functional groups (azides, tetrazoles) yield N_2 on combustion and are not detected
	Easily integrated online with separation systems	
MS	Excellent detection sensitivity	Response factors depend upon compound ionization efficiencies and sample matrix effects
	Determination of exact mass enables compound identification	
	Easily integrated online with separation systems	Limited mass resolution for high MW compounds
NMR	Universal response to specific nuclei, independent of compound structures	Spectral assignments required; complex data interpretation
	Can provide quantitative purity measurements when used with internal standards	Peak assignments difficult in the presence of impurities
	Can provide detailed structural information for compounds	Difficult to integrate online with separation systems

have been developed to analyze discrete compounds and mixtures of compounds.

6.2   Analysis of discrete compound libraries

6.2.1   MS analysis

The validation of libraries from high-throughput, parallel, solid-phase synthesis is achieved mostly by cleaving compounds from solid support. Therefore, the samples for analysis are in a format similar to libraries made by high-throughput, parallel, solution-phase synthesis. In contrast to split libraries, discrete (parallel-synthesis) compound libraries are well suited for high-throughput MS analysis. First, multiple parallel synthesis should generate a single end product with a known molecular weight (MW) in a specific well location. Second, these end products are delivered for analysis in solution and in standardized plate format suitable for automated laboratory handling, e.g., 96-well plates. Third, the sensitivity is rarely a problem for MS from the amount of compounds generated by combinatorial synthesis (Schriemer et al. 1998; Yates et al. 2001).

High-throughput electrospray ionization mass spectrometry (ESI-MS) analysis for molecular weight determination of synthetic peptides including characterization of byproducts arising from synthesis has been described (Smart et al. 1996). The data from electrospray mass spectrometry are processed with an algorithm that calculates the contribution of the target peptide and each of the identifiable contaminants to the total ionizable material in a sample of synthetic peptide. All essential data are obtained in less than three minutes per sample. In this work, the advantage of this technique was the rapid analysis of the many hundreds of peptides/peptidomimetics required in systematic quantitative structure-activity studies.

The ESI LC/MS interfaces developed in the early 1990s proved to be timely technologies for rapidly providing molecular weight information for diverse samples. The only challenge is the dauntingly high throughput needed (e.g., 50 to 100 thousand analyses per year) and the resulting data interpretation problem.

An automated open-access flow-injection analysis (FIA) MS system has been developed for the QC of combinatorial libraries (Hegy et al. 1996). Using this system, an analysis of compounds from a 96-well plate could be completed in three hours in 1996. Three tasks are automatically accomplished in each run: the confirmation of the presence of the expected compound, the detection of any intersample carryover, and the automated reporting of results to the chemist's bench.

Enhancements to the FIA-MS system (Gorlach et al. 1998) were specifically designed to help the combinatorial chemist improve the end product purity by pinpointing synthetic inefficiencies. A 3-D spectral map of the entire combinatorial chemistry rack assists in reaction optimization by color highlighting contamination ions that are detrimental to overall purity.

The size of some large libraries originally deterred full characterization in a reasonable time period. In this situation, analysis of a statistically relevant subset of the combinatorial library by FIA-MS was done prior to biological testing. However, the recent development of subminute duty cycles for FIA-MS can now support a complete or near complete analytical survey of even large libraries.

6.2.2 MS-guided purification

Although the parallel synthesis (one compound per well) strategy eliminates the deconvolution steps, which is necessary when pooled libraries are screened, parallel synthesis products are usually screened as crude mixtures because purification slows the process of lead discovery. However, screening crude products can produce misleading results due to active side products, unreacted material, or synergistic interaction of multiple components. Data from "active" wells can be evaluated only following resynthesis and purification, and data from "inactive" wells may be unreliable. Although a total purification approach may not be necessary for lead discovery, screening pure compounds is the only sure route to immediate, reliable structure-activity relationships. Therefore, there is a clear advantage in purifying individual products obtained from parallel synthesis. Such an effort has recently been reported (Kiplinger et al. 1998). In the approach using serial purification of samples made from parallel synthesis, speed is a key issue. In this work, a 10 mm x 5 mm C-18 HPLC column was used. Because of the moderate backpressures generated, operation at elevated flows is feasible. The separations were performed at a flow rate of 5 ml/min. While use of shorter HPLC column length sacrifices theoretical plates, the vast majority of separations involving crude reaction mixtures do not require high efficiency, and the reduction in column length with an increased flow allows for sharply increased throughput. This purification procedure produces material of acceptable purity with a good recovery rate, and product compounds are readily quantified by weighing.

Rapid reversed-phase analytical and preparative HPLC methods have been developed for application to parallel synthesis libraries (Weller et al. 1997). Gradient methods, short columns, and high flow rates allow analyses of over 300 compounds per day on a single system, or purification of up to 200 compounds per day on a single preparative system.

Among other multiple-column HPLC systems, an automated parallel analytical/preparative HPLC/MS workstation (Analyt/PrepLCMS) has been developed to increase the analysis speed for characterizing and purifying combinatorial libraries (Zeng and Kassel 1998). The system incorporates two columns operated in parallel for both LC/MS analysis and preparative LC/MS purification. This system has a multiple sprayer IonSpray interface to support flows from multiple columns and a valving configuration, that is under complete software control for delivering the crude samples to the two HPLC columns from a single autosampler. By incorporating fast chromatography and mass-based fraction collection on the new parallel Analyt/PrepLCMS system, more than 200 compounds can be characterized per instrument per day, with the same amount per night.

An even higher throughput analysis of compounds synthesized by parallel synthesis has been reported recently (Wang et al. 1998). Compounds can be processed eight at a time using this eight-probe autosampler coupled with electrospray ionization. There was no obvious advantage of one configuration over the other.

6.2.3 High-throughput NMR

There are two reasons to focus on the analysis of solution-phase samples: first, as stated in a previous section, most drug discovery programs perform the critical bioassay studies on solution state (cleaved) compounds, and second, the automated manipulation of liquid samples is much easier (and more reliable) than the automated manipulation of heterogeneous SPOS resin slurries.

Although HT MS has been applied to characterize compounds from combinatorial libraries, NMR has not been used in a similar fashion. There are four reasons for this:

1. The samples are normally dissolved in protonated solvents that generate large background signals that complicate the NMR spectra.
2. The volumes of solute are often too small for conventional NMR spectroscopy (< 500 µl). The extreme case is a "one-bead-one-compound" library.
3. The samples are often in a 96-well microtiter plate format that cannot be used directly by a regular NMR probe.
4. Conventional NMR analysis is too slow.

HT NMR will be possible only when these problems are solved. The solvent interference has been overcome by a solvent suppression method called "WET." WET uses a series of four frequency-selective (shaped rf) pulses, each having simultaneous ^{13}C decoupling and each followed by a pulsed field gradient (PFG); the resulting four-element pattern has been optimized to create a highly effective solvent suppression pulse sequence element that has been incorporated into a number of 1D- and 2D-NMR sequences (Smallcombe et al. 1995).

The problem of the small sample volume has been addressed by the development of highly sensitive on-flow LC-NMR probes designed for small volume samples such as HPLC fractions. The sensitivity of such probes allows them to detect nanomole amounts of samples. These developments have elevated LC-NMR into a high-performance technique. This technique is shown as technique 1 in Figure 6.1 as an LC-NMR setup. Although it is referred to as "on-flow" NMR, the solvent flow may always be temporarily paused to trap the sample in the detector coil during acquisition to allow averaging to be used to improve sensitivity.

The problems of 96-well plate format and low NMR throughput have been solved by incorporating an HPLC autosampler at the front end of the LC-NMR. This creates an "autosampled LC-NMR" (technique 2 in Figure 6.1). The critical aspect in these two formats is the delivering of a sample to the NMR probe by using a flowing liquid.

Removing the HPLC column simplifies the operation. The autosampler injects an aliquot of sample from a 96-well plate. The solvent will sweep it through the tubing and into the NMR probe (technique 3 in Figure 6.1). The signal can be averaged in the NMR coil. This process, when combined with certain scanning techniques and optional locking and shimming, is quite fast. A ^{1}H NMR spectrum requires one minute of signal averaging (32 scans), and the entire "deliver/acquire/flat-out" process for each sample may take three minutes.

The system in technique 3 can be further simplified by removing the HPLC pump and the mobile phase (technique 4 in Figure 6.1). In this direct-injection system, the autosampler takes a sample aliquot from

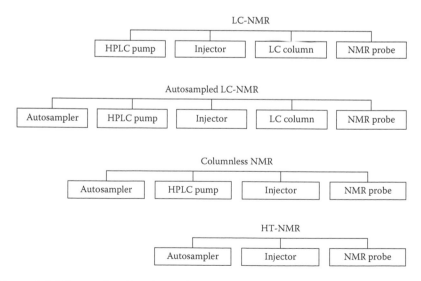

Figure 6.1 Scheme showing the evolution of HT-NMR spectroscopy.

a 96-well plate (100–300 µl) and, by using the injector, loads it directly into the detector cell of the NMR probe—without the use of any mobile phase or dilution. After completion, the process is reversed to return the intact sample directly back into the microtiter plate. This technique was reviewed by Keifer (1997, 1998). Examples of spectra recorded using techniques 3 and 4 are shown in Figure 6.2.

Techniques 1 and 2 have the advantage of being able to provide chromatographic separation for mixtures, but only at a disadvantage in increase in analysis time. Techniques 3 and 4 use the opposite approach—no chromatographic separation—so the analysis is fast. Technique 4 produces higher sensitivity than "columnless NMR" because there is no dilution. Because a sample can be analyzed in less than three minutes, this technique can be viewed as the first generation of HT NMR spectroscopy. This HT NMR spectra acquisition will make the spectra interpretation the next bottleneck. However, recently developed NMR techniques, such as proton chemical shift spectra from filtered J-spectra, reference deconvolution, spectral calculations and automated spectral analysis, offer the potential that a fully automated real-time microtiter plate NMR acquisition and data analysis system for combinatorial chemistry libraries might some day be developed.

6.3 Analysis of pooled libraries

Combinatorial synthesis using split and pool protocol (Furka et al. 1988, 1991; Lam et al. 1991; Houghten et al. 1991) produces one-bead-one-compound libraries. Although discrete compounds are actually made at the single-bead level, the analysis of a mixture is encountered when the library is to be analyzed as a pool.

Characterization of a pooled library is not trivial. Any destructive analytical methods will not be useful because, theoretically, only one bead carries one unique compound. The huge number of compounds in the final library made by this method precludes the possibility of a full characterization. The focus has to be on the synthetic route optimization and the characterization of small-size trial libraries.

The amounts of peptides obtained from a single bead were sufficient to yield abundant quasi-molecular ions in MALDI-TOF and MALDI-FTMS spectra. CID experiments were performed and yielded satisfactory spectra. Peptides from a single bead can be cleaved directly on the MALDI probe to reduce sample preparation times and handling (Franz et al. 2003).

6.3.1 Theoretical calculation of mass distribution

The feasibility of probing combinatorial library diversity by MS is addressed theoretically using computer-generated mass distributions of

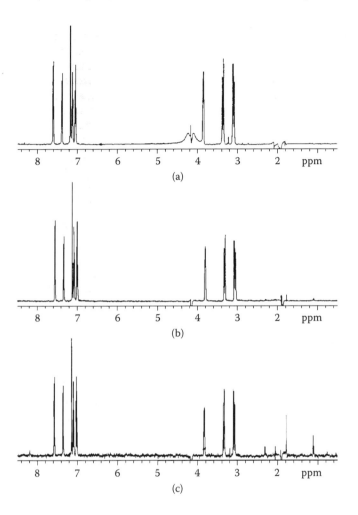

Figure 6.2 **(Opposite)**

combinatorial synthesized peptide libraries containing between two and seven amino acids (Demirev and Zubarev 1997). The behavior of several "global" (integral) parameters of such mass distributions—mass centroid, dispersion, skewness, and kurtosis—was studied. The centroid and dispersion are shown to carry information that may characterize the completeness of the synthetic effort.

Based on determining global (integral) as well as local characteristics of the overall mass distributions of all individual library components, the implementation of the computer simulation seems feasible for libraries containing from several hundred up to several thousand individual components. Such an approach can also be useful for the MS monitoring of

Figure 6.2 Direct-injection and "on-flow" ¹H spectra acquired on tryptophan solution (1 mg/mL). WET solvent suppression was used for both CH_3CN resonance (at 1.95 ppm) and for the $D_2O/HOD/H_2O$ resonance (at 4.18 ppm); WET also selectively suppressed the ¹³C satellites of acetonitrile. The top spectrum (a) was acquired with HT NMR, with the tryptophan dissolved in a fully protonated solvent mixture (1 mg/mL tryptophan dissolved in 50:50 $CH_3CN:H_2O$). This is possible because WET suppression does not require the NMR to maintain a ²H "lock" for stability of the magnetic field (so a deuterated solvent is not required) and WET is fully capable of high-quality H_2O-signal suppression. The middle spectrum (b) was also acquired by HT NMR to inject a tryptophan solution directly into the probe (and return it directly to the sample container when finished), but a different tryptophan solution (1 mg/mL in 50:50 $CH_3CN:D_2O$) was used. The bottom spectrum (c) was acquired using "columnless" LC-NMR, with 100 µL of tryptophan solution (1 mg/mL in 50:50 $CH_3CN:D_2O$). In "columnless" LC-NMR the mobile-phase sweeps the sample from the injector loop into the NMR probe for detection, then rinses out the NMR probe at the end of signal averaging; analogous to a "stopped-flow" version of a "carrier gas" in GC. All spectra were acquired at 500 MHz, using 1-min acquisitions (32 scans each), without any form of chromatographic separation. (Reproduced with permission from *Drug Disc. Today*, 2, 468478, 1997. Copyright© 1997. Elsevier Science Ltd.)

the diversity of reaction products containing around 10^3 individual members at each subsequent step along the combinatorial synthetic pathway. Appropriate statistical criteria still have to be formulated in order to prescribe the adequate sampling rate, resulting in a predetermined tolerance limit within a given confidence interval.

Molecules have different ESI-MS sensitivities depending on the functional groups present. It is beneficial to examine both a positive ion and negative ion spectrum to obtain a more realistic picture of the components. In summary, for larger libraries where mass resolution of each individual component is prohibitive, one can compare a calculated theoretical mass distribution with the measured spectrum to gain some insight. The shape and position of the unresolved distribution provide useful information regarding the success of the synthesis.

6.3.2 MS analysis

In combinatorial synthesis, the formation of byproducts during the synthesis cannot be completely avoided. Major reasons for the occurrence of byproducts are, for example, the incomplete coupling or deprotection steps and the undesirable oxidation by the oxygen in the air. Because the reaction is applied not to a single compound but to an entire population of different substrate molecules, there is always the potential of encountering some selectivity toward the generation of certain molecules at the expense of others. Impurities or missing components in library mixtures used in biological assays can lead to misinterpretation of the obtained results. Therefore, quality control of pooled libraries must accomplish two tasks:

1. To confirm whether all of the expected compounds were produced during the combinatorial synthesis.
2. To identify byproducts.

The analysis of compound mixture library is not so straightforward. Because of severe peak overlaps, techniques such as NMR, IR, UV, and fluorescence spectroscopy cannot provide meaningful structural information. However, ESI-MS and matrix-assisted laser desorption/time of flight mass spectrometry (MALDI-TOF-MS) produce essentially fragmentation-free spectra with a pseudomolecular ion peak for every component. They have been widely used for mixture analysis.

Here is a novel qualitative approach for the screening of a small library of compounds using MALDI-TOF-MS and HPLC-ESI-MS/MS. This method does not guarantee the provision of complete information about all binding components principally capable of binding, but it certainly ensures the detection and identification of the binding components having concentrations above the detection limit and/or possessing relatively high affinities to the target protein. In the screening of complex ligand mixtures, MALDI-TOF-MS and HPLC-MS/MS are rapid and highly effective (Mathur et al. 2003). The approach of imaging time-of-flight secondary ion mass spectrometry (TOF-SIMS) acquiring mass spectra from arrays of polymer resin particles can increase the analysis speed (see Figure 6.3). The speed is determined by the instrumental parameters and the density of the arrays (Xu et al. 2003).

A general limitation in the use of common ionization techniques for the analysis of mixtures is that ionization of some components in a mixture may be partially or completely suppressed, even though the same compounds when analyzed alone give intense molecular ions.

ESI-MS has been used to establish that a sizable fraction of the expected compounds are produced in the synthesis of compound libraries (Dunayevskiy et al. 1995). Experiments were designed to determine whether the combination of a xanthene acid chloride core with a set of amines results in the formation of a mixture in which all of the expected compounds are present above a certain concentration threshold. Direct observation of 65,341 compounds in the final library was out of the question, and even a library of 136 compounds (only four building blocks) appeared daunting in size. Therefore, smaller yet representative trial libraries were constructed. In a library of 55 expected compounds, 43 (78%) were detected. Missing were predominantly those compounds that contained the arginine-p-benzylamide building block (Dunayevskiy et al. 1996).

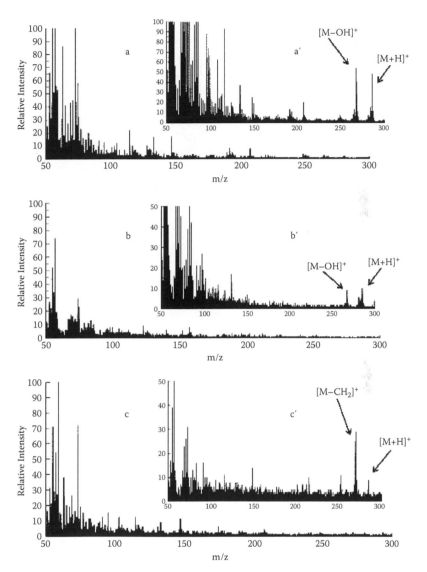

Figure 6.3 SIMS spectra of PS resins functionalized with stearic acid through various linkers-Wang, thermally labile, and photosensitive linkers before (a, b, c) and after (a′, b′, c′) linker cleavage using the methods described in the Experimental Section. In all cases, no protonated molecule [M + H]$^+$ of stearic acid (m/z 285.28) is observed before linker cleavage, but after the molecules are clipped from the linker, ions of [M + H]$^+$ (m/z 285.28), [M - OH]$^+$ (m/z 267.26), or [M - CH$_2$]$^+$ (m/z 270.26) are detected in the SIMS experiments. (a, a′) Stearic acid anchored on the PS beads through acid-sensitive Wang linker. (b, b′) Stearic acid on thermally labile linker. (c, c′) Stearic acid on hydroxyethyl-photosensitive linker. (Reproduced with permission from *Anal. Chem.* 75, 6155–6162, 2003. Copyright© 2003, American Chemical Society.)

Even though the analysis of mixtures with a diversity of 10^4 to 10^5 is not yet possible, the preparation of specifically designed model libraries and their investigation by ESI mass spectrometry has proven to be a useful technique for the analysis of mixtures of structurally related compounds. With the mass-spectrometric data in hand, one is confident that the synthetic methodology produced highly diverse libraries of well-defined composition, and one is able to put faith in the results of subsequent screening assays.

Because MS identifies the synthetic compounds based on molecular mass, the presence of a large number of isobaric isomers in a mixture is the major problem. In a 48-component octapeptide library with three variable amino acids, only 16 different mass values were detected (Metzger et al. 1994). Whether it is a small pool or a large pool of compounds, no structural information can be extracted if isobaric problems exist. Coupling MS to other separation techniques or the use of high-resolution MS or tandem MS are required to resolve isobaric mixtures in small library analysis.

For a compound mixture, mass peaks within the m/z range for the quasi-molecular ions of the lightest and heaviest peptides of the mixture are expected. The presence of peaks at lower or higher m/z values immediately indicates byproducts (for example, deletion peptides or peptides containing uncleaved side-chain protecting groups). In this way, it is possible to rapidly reveal byproducts even in libraries containing billions of peptides.

Compound mixtures can be conveniently introduced into the ESI source via an autosampler (60 samples per hour). Experimental data are then compared with the characteristic mass and intensity distribution of the library or sublibrary with the interactive help of a computer program that predicts mass distribution of a compound library. This kind of program, as discussed in Section 6.3.1, is based on a linear relationship between the number of isobaric peptides and intensity. Byproducts can be revealed very quickly. Structural information on byproducts can be obtained by using tandem mass spectrometry with daughter ion, parent ion, or constant neutral loss scans. For smaller libraries, online HPLC-MS helps distinguish isobaric peptides or identify individual components in the mixture.

ESI-MS, tandem MS, and online RP-HPLC-ESI-MS were used to evaluate the composition and purity of three different aryl ether mixtures consisting of 10 and 45 aryl ethers synthesized on solid support by Williamson etherification (Haap et al. 1997). The library features two potential pharmacophores connected with three different spacers and serves as a models for a detailed component analysis. When libraries were made by two synthetic methods, the split-and-pool method and the premix method, different mass distributions in the ESI-MS spectra were observed. Components that could not be detected by the ESI-MS method were found by MS/MS experiments. Parent ion and constant neutral loss scans allowed the identification of components with common structural features.

The incorporation of the high-throughput LC-MS method (Daley et al. 1997) further increased the efficiency of the analysis of pooled compounds. For example, if 10 96-well plates are logged in, and each well contains 10 components, then 9600 compounds can be confirmed in 16 hours.

A quality-control analysis of an encoded combinatorial library produced by split synthesis was reported (Lewis et al. 1998). Because the library was made to have at least three copies of each component, the randomly selected individual beads were first photocleaved for analysis with capillary HPLC/ESI-MS analysis. Then, the dialkylamine tags used to record the synthesis steps were chemically cleaved from the bead and decoded by HPLC with fluorescence detection. In this way, they could compare the expected products from the decoding results with the detected products by LC-MS. These results confirmed the success of the library synthesis and allowed chemists to carry this library confidently into high-throughput screening for biological activity.

In TLC/MALDI-TOF-MS analysis, the reaction mixture is separated by TLC and the analyte is extracted and analyzed by MALDI-TOF-MS, giving the mass of the compound in a relatively short time. The compounds are separated on TLC plates, scraped off, and extracted and analyzed by MALDI-TOF-MS using standard procedures. The methodology has been applied to the analysis of several classes of compounds from a variety of reaction mixtures (Hilaire et al. 1998).

There are some advantages of the HPLC/MS-TOF technique over standard LC/MS. One important feature is the separation of mixture components into distinct charge-state families. This separation makes it possible to obtain integrated mass spectra for specific families of ion charge states for unambiguous mass assignment. In addition, many isomeric library peptides that have identical LC retention times can be separated on the basis of differences in their gas-phase mobilities in the $t_r[t_D(t_f)]$ data (t_D is drift time, t_f is flight time) (Barnes et al. 2002).

6.3.3   High-resolution MS

Many times, however, two or more components in a mixture cannot be unambiguously identified because they have nearly identical masses that cannot be distinguished by conventional mass spectrometers. For example, the substitution of lysine for glutamine in a peptide yields a molecular weight difference of 0.036 mass units.

Frontal immunoaffinity chromatography with mass spectrometric detection can be used to rapidly isolate, screen, and identify various active compounds in the extract of a plant in a single step and to sequence them by their activities. The method can be applied to find new leading

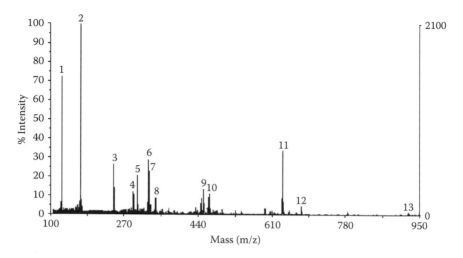

Figure 6.4 Mass spectrogram of the extract solution. The 13 labeled peaks are major compounds in the extract. (Reproduced with permission from *Anal. Chem.* 75, 3994–3998, 2003. Copyright© 2003, American Chemical Society.)

compounds from nature or from a man-made combinatorial library that have different structure types, or to find substitutes for a synthetic active compound that has high toxicity (See Figure 6.4, Luo et al. 2003).

ESI Fourier transform mass spectrometry (FTMS) can be used to address such common challenges in characterizing complex mixtures (Winger and Campana 1996). For example, a phenylalanine-substituted octapeptide ion (F) at M/Z 964 had a measured mass (963.5346) that was in error by 17 ppm.

The utility of Fourier transform ion cyclotron resonance (FTICR) mass spectrometry for analysis of selected small-peptide combinatorial libraries containing 10^2 to 10^4 individual components has been demonstrated (Nawrocki et al. 1996). The FTICR technique offers many advantages over conventional mass spectrometric methods. These include ultrahigh mass resolution, high mass accuracy, multistage tandem mass spectrometry (MS^n), and the ability to trap ions for extended periods of time. As libraries become larger and more diverse, it will be increasingly difficult to separate individual peaks with the same nominal mass-to-charge ratio. However, with the high mass resolution afforded by the FTICR technique, it is possible to mass resolve an individual compound in isobaric mixtures. The high-mass-resolution spectra obtained by FTICR MS can reveal details about libraries that would otherwise be obscured in data from lower resolution mass spectrometric techniques. A study of pooled libraries with 36, 78, and 120 components showed that 70–80% of the library components can be identified without separation (Fang et al. 1998). Furthermore, the FTICR technique can be combined with collisionally activated dissociation (CAD) or infrared multiphoton dissociation (IRMPD) of isolated species to yield a range of information about the structures of individual components. The mass resolution can further be enhanced because the mass resolution increases linearly with the field strength of the magnet.

Multidimensional automated chromatographic techniques coupled to electrospray ionization mass spectrometry facilitate the selection process and provide maximum characterization information in a single screening run. Advances in the parallel syntheses and automated high-throughput screening methods for compounds as potential drug candidates have increased the throughput of drug discovery (Hsieh et al. 1996).

6.3.4   Tandem MS

Small libraries are often constructed as a model study for final synthesis of a large library. The quality of such libraries can be assessed by ESI-MS and tandem MS. A study of xanthene derivatives clearly shows the complementary roles of positive and negative ion ESI analysis (Dunayevskiy et al. 1995). Figure 6.5 compares the mass spectrum of the same set of compounds differing in acidic groups on the molecule in positive and negative ion mode. The compound lacking acidic residues, such as TrpMe/X/ TrpMe, disappears completely in the negative ion spectrum. In general, the presence of those compounds, which were obscured in positive (negative) ion mass spectra, could be verified by negative (positive) ion detection. Ion mobility techniques can provide a detailed characterization of complicated mixtures. These methods appear to be suitable for analysis of the wide range of sequence, structural, and stereo isomers—as long as such isomers display differences in collision cross-section (Barnes et. al. 2000).

A tandem mass spectrometer consists of an ion source, two analyzers separated by a fragmentation region, and an ion detector. The ionization of the mixture produces ions characteristic of the individual components. The parent ion undergoes collisionally activated dissociation (CAD) through collisions with neutral gas molecules in the fragmentation region to yield various daughter ions. Subsequent mass analysis of the daughter ions by the second mass analyzer permits identification of the separated components.

Although the specific operating mode chosen for a particular analysis will depend on the information desired, the three most common MS/MS operating modes of the tandem mass spectrometer are the daughter ion scan, the parent ion scan, and the neutral loss scan. A daughter ion scan consists of selecting a parent ion characteristic of the analyte by the first mass analyzer, fragmenting it by CAD in the fragmentation region, and scanning the second mass analyzer to obtain a daughter mass spectrum. Analogous to a normal mass spectrum, the daughter mass spectrum can be used for identification of an analyte by standard mass spectral

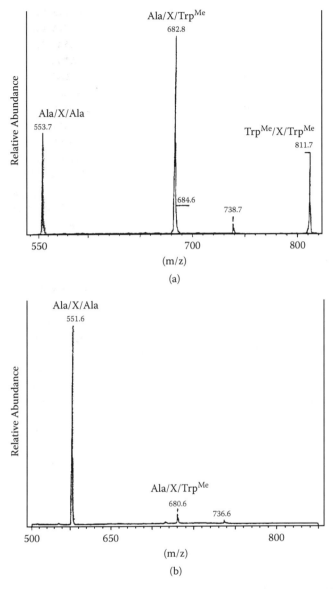

Figure 6.5 Mass spectra of the mixture of the mixture of $Trp^{Ma}/X/Trp^{Me}$, Ala/X/ Trp^{Me}, and Ala/ X/Ala. (a) Positive ion and (b) negative ion ESI. (Reproduced with permission from *Anal. Chem.* 67, 2909, 1995. Copyright© 1995, American Chemical Society.)

interpretation. In the parent ion scan mode, a specific daughter ion is selected with the second mass analyzer, and the first mass analyzer is scanned over a specific mass range, selecting parent ions of different m/z that fragment to yield the specific daughter ion. This mode can be used to screen for a class of compounds that fragment to yield a common substructure. In a neutral loss scan, both mass analyzers are scanned with constant difference in mass. The resulting neutral loss spectrum contains the daughter ions that arise from the loss of a specific neutral fragment from the parent ion. This mode is useful for screening for a class of compounds characterized by a common fragmentation pathway.

The accurate mass and MS/MS data can be rapidly and easily obtained by an automated MALDI-FT-MS procedure for multiple samples. Multiple ionization techniques can also be used simultaneously in order to introduce sample and mass reference ions into the FTICR cell. In this way 20 samples from a combinatorial libraries were analyzed automatically in one hour, with errors of less than 5 ppm for the masses of both parent and fragment (Tutko et al. 1998).

The amounts of peptides obtained from a single bead were sufficient to yield abundant quasi-molecular ions in MALDI-TOF and MALDI-FT-MS spectra. CID experiments were performed and yielded satisfactory spectra. Peptides from single beads can be cleaved directly on the MALDI probe to reduce sample preparation times and handling (Franz et al. 2003).

A "basket in a basket" method based on a multistage accurate mass spectrometric (MAMS) technique was developed and demonstrated (Wu 1998) by obtaining a unique elemental composition of a compound (with a molecular weight of 517) from combinatorial synthesis. The accurate masses for the parent and the fragment ions were obtained with up to five stages of MAMS using FTICR-MS.

If the higher mass ions are dissociated into smaller fragments, it is possible to use the uniquely defined compositions of these fragment ions to "reconstruct" the elemental compositions of the parent ions. Not only does the composition of the smaller fragments serve to determine those for the larger fragments, but also the composition of these large fragments (larger rings) set the upper limit for the next generation of ions. This is similar to putting a smaller basket into a bigger one—a mutual confinement would occur so that only limited shapes of the baskets would fit into each other. Clearly, such a "basket in a basket" approach can be best implemented by combining multistage fragmentation techniques (MS^n) with accurate mass measurements, that is, by using multistage accurate mass spectrometry (MAMS). The "basket in a basket" approach also greatly reduces mass accuracy requirements because the original larger ions were dissociated into smaller fragments, which could be determined using only a moderate mass accuracy.

The coupling of a reliable nanoelectrospray ionization technique with MAMS is important for its practical usage. In general, the amount of sample available for analytical work is limited, at least in the lead generation stage of combinatorial chemistry. There have been many cases where biological activities came from unexpected products of the combinatorial synthesis. In such cases, structural determination becomes extremely important, while the amount of sample available for analysis is usually small. In the experiments using MS^2 to MS^5 (Wu 1998), a total of 3 µl of the solution was loaded into the flow injection source, that is, the amount of the analyte compound injected was ~30 pmol. Such experiments would be impossible without the nanoelectrospray technique.

The micellar selectrokinetic chromatography (MEKC) method is a nice alternative to HPLC when the need to reduce the use of solvents and sample is a concern. In addition, it is an extra tool for assessing the purity of combinatorial chemistry libraries (Simms et al. 2001).

6.3.5 Liquid chromatography (LC)/MS and capillary electrophoresis (CE)/MS

6.3.5.1 LC-MS

The combination of HPLC with MS divides a complicated mixture library into many small mixture groups, and the identification of compounds becomes easier (Metzger et al. 1993). For an octapeptide library of 48 components, individual components cannot be identified from the MS spectrum. A LC-MS measurement was performed; the data are shown in Figure 6.6. Panel (a) displays the HPLC trace of all molecular ions, and panel (b) shows only components with the mass value of 1040.6. Owing to the isobaric isomers, at least eight components have the mass value of 1040.6. In each fraction at various retention times, multiple compounds were detected. This example demonstrates how the complexity of the mixture library can be resolved/reduced using LC-MS techniques.

Online sample preparation can increase the efficiency of high throughput LC/MS analysis. In addition to reducing the probability of contamination that results from employing a manual preparation process, it can also increase sample recovery in the special case where solvent requirements for sample preparation and chromatography are incompatible (Tang et al. 2001).

LC/ESI-MS with selected-ion monitoring makes possible the visualization and positive identification of a family of isobaric ions from libraries comprising hundreds of peptides (Lambert et al. 1997). For larger libraries (up to 105 peptides), a given peptide, known to be unique at a given *m/z*, was detected, but the full characterization of others at other *m/z* values

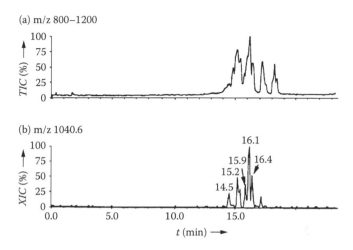

Figure 6.6 HPLC-MS of a peptide library. (a) Reconstructed total ion chromatogram (*m/z* 800-1200) and (b) ion chromatogram of *m/z* 1040.6 corresponding to the monoisotopic $(M+H)^+$ ions of eight isobaric peptides of the mixture (TIC: total ion current; XIC: extracted ion current). (Reproduced with permission from *Angew. Chem. Int. Ed. Engl.* 32, 894–896, 1993. Copyright© 1993, WILEY-VCH Verlag GmbH.)

from the library is limited mainly by LC separation efficiency. However, this approach offers new possibilities for evaluating the completeness of combinatorial libraries.

6.3.5.2 CE-MS

Capillary electrophoresis (CE) generates rapid, high-resolution separations based on differences in the electrophoretic mobility of charged species in an electric field in small-diameter fused-silica capillaries.

Reliable high-throughput analyses can be performed based on CE or MEKC. In this way, the throughput of CE was improved 100-fold. The use of two internal standards greatly improved the reproducibility of multiplexed CE. RSDs (relative standard deviation) for the normalized migration times become comparable to those for GC and for LC (Xue et al. 1999).

Application of CE before mass spectrometric detection provides a fast and efficient separation of library components for their subsequent characterization by ESI-MS. When voltage is applied across the capillary, electro-osmotic flow (EOF) moves the sample from the positive end (anode) to the negative end (cathode). EOF becomes significant at pH values >5 and drives the analytes toward the cathode, regardless of their charge. The mobility of the library compounds and, therefore, their migration times are determined by the difference between the mobility of the EOF and the electrophoretic mobility of the species. Buffer composition, pH, and ionic strength can be varied to optimize electrophoretic separation. The best electrophoretic condition could be chosen to separate a complex mixture into several groups of analytes, each migrating according to its charge. Within each such group, one can expect that significantly fewer isobaric compounds compared with the unresolved library mass spectrum. A small model library of xanthene derivatives (Diagram 6.1) was analyzed by CE-ESI MS to test the possibility of making a large library based on a xanthene core (Dunayevskiy et al. 1996). Figure 6.7 shows the electropherogram of the library and extracted ion electropherograms with selected ion current. In this characterization, 160 of 171 compounds in the library were identified.

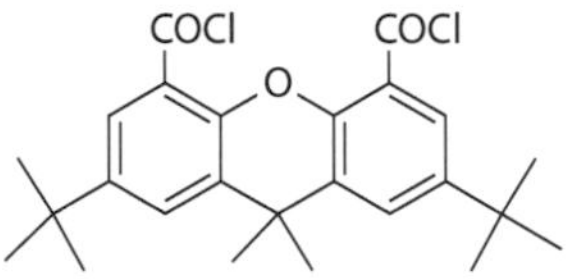

	Amino Acid	Protected Reagent Used
1.	(Gly)-OMe	Glycine-methyl ester
2.	(Ala)	L-alanine-tert-butyl ester
3.	(Ser)	O-tert-butyl-L-serine-tert-butyl ester
4.	(Pro)	L-proline-tert-butyl ester
5.	(Val)	L-valine-tert-butyl ester
6.	(Leu)	L-leucine-tert-butyl ester
7.	(Ile)	L-isoleucine-tert-butyl ester
8.	(Asn)	L-asparagine-tert-butyl ester
9.	(Asp)	L-aspartic acid-?tert-butyl-a-tert-butyl ester
10.	(Thr)-OMe	O-tert-butyl-L-threonine-methyl ester
11.	(Glu)	L-glutamic acid-γ-tert-butyl-α-tert-butyl ester
12.	(His)	N^{im}-trity-L-hiscidine
13.	(Lys)-OMe	N^E-Boc-L-lysine-methyl ester
14.	(Met)-OMe	L-methionine-methyl ester
15.	(Phe)	L-phenylalanine-tert-butyl ester
16.	(Arg)	N^g-4-methoxy-2,3,6-trimethylbenzene-sulfonyl-L-arginine
17.	(Trp)-OMe	L-tryptophane-mentyl ester
18.	(Tyr)-OMe	O-tert-butyl-L-tyrosine-methyl ester

Diagram 6.1

With reversed-phase, high-performance liquid chromatography (HPLC), the separation of physicochemically diverse mixtures presents a significant challenge. The separation in CE is based on differences in charge and size and provides significantly higher efficiencies than does HPLC. The applicability of CE as a separation tool for N-(substituted)-

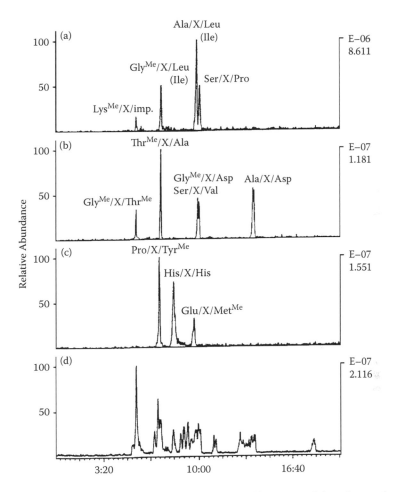

Figure 6.7 CE-MS electropherogram of a library with extracted ion electropherograms. (a) m/z = 595.5; (b) m/z = 597.5; (c) m/z = 685.5; (d) Total ion current electropherogram. (Reproduced with permission from *Proc. Natl. Acad. Sci. U.S.A.* 93, 6155, 1996. Copyright© 1996, National Academy of Sciences, USA.)

glycine-peptoid combinatorial mixtures was demonstrated (Robinson et al. 1998). A multiple-background electrolyte additive strategy provided enhanced selectivity over the additives used individually. An ion-pairing agent, 1-heptane sulphonic acid (HSA), presumably reduced the propensity to form strong intramolecular interactions between the peptoids and disrupted electrostatic interactions with the capillary wall. Methyl-β-cyclodextran was used as a partitioning component in the separation to aid in the resolution of the more hydrophobic peptoids. The addition of organic solvents to the background electrolyte, alone and in conjunction with HSA, has been shown to enhance the separation of peptoids.

CE offers potential advantages over traditional HPLC for the analysis of complex peptoid mixtures. Only nanoliters of the sample are injected for each analysis, and as little as 5 µL of the sample is needed for a reproducible injection. This can be a tremendous advantage when extremely small amounts of the sample are available, as is typical with combinatorial chemistry. Separations are inherently based on charge-to-size ratios, and partitioning can be incorporated, providing greater flexibility in developing a separation.

6.3.6   NMR methods for mixture analysis

In order to achieve purification and structural identification simultaneously for compound mixtures from either the natural sources or from split synthesis, HPLC-NMR has been studied as a useful analytical tool.

In HPLC-NMR operations, it is possible to achieve ^{1}H NMR spectroscopic detection of HPLC elution in three modes of operation:

1. Direct on-flow detection of NMR spectra in real time is possible in favorable circumstances.
2. Stopped-flow analysis allows the full range of NMR experiments to be performed on a defined chromatographic fraction.
3. It is possible to collect fractions of the HPLC eluate into capillary loops for later off-line full NMR spectroscopic evaluation.

A model system based on a mixture of 27 tripeptides combinations of alanine, methionine, and tyrosine as the C-terminal amides has been studied (Lindon et al. 1995). The ^{1}H NMR spectra were obtained in real time using on-flow HPLC-NMR spectroscopy at 600 MHz, and 21 of the 27 peptides were identified on the basis of chemical shifts and coupling constants, including the use of diagnostic values derived from ^{1}H NMR spectra and ^{1}H-^{13}C and ^{1}H-^{15}N HMQC studies on authentic AAA-OH and YYY-OH.

A model four-component mixture of aromatic ring positional isomers was studied (Chin et al. 1998). These isomers cannot be identified by MS without time-consuming MS/MS studies. The mixture (100 µg) was first analyzed by stopped-flow HPLC-NMR. For these data, 128 scans were more than sufficient to obtain suitable S/N. The structure determination for each compound is obtained by consideration of the chemical shifts and the coupling patterns. In an on-flow HPLC/NMR analysis, approximately 200 µg of four-component mixture was used.

An NMR strategy for mixture analysis without physical separation of the components was reported (Lin and Shapiro 1997). The method is based on the combination of HMBC with TOCSY. As in the case with NMR assignments for peptides and small proteins, a structure verification rather than a *de novo* structure determination is usually involved in

combinatorial chemistry problems. The strategy used for structure verification is similar to that employed to perform structure assignment in
peptides and small proteins. A TOCSY spectrum was initially obtained to
establish the individual spin system connectivity in a manner analogous
to the peptide example. However, in general, the total spin connectivity
cannot be accomplished, owing to nonreporting centers. For most organic
compounds, the spin systems are broken into several fragments, which
must be connected to regain molecular identity. Further connectivity can
be made by HMBC. The HMBC experiment connects protons to carbons
via two or three bond couplings. This allows the connection of protonated
carbons to nonprotonated carbon centers, thereby allowing the spin systems derived from the TOCSY experiment to be linked. Applying this
principle, the TOCSY spectrum of the mixture of six esters was collected.
Following chemical shift information, the spin fragmentation of organic
groups can be obtained. The following HMBC data allow the connection
of separate molecular groupings to identify molecules. If the ^{13}C chemical
shifts for the common connecting groups are insufficiently resolved to
allow assignment, the diffusion-encoded NMR methods can be used to
aid the structural assignment (Lin and Shapiro 1996).

6.4   HPLC for analyzing compounds from discrete and mixture libraries

An important early step in the characterization of the compound is to
obtain an accurate concentration of the compound in a solution in order to
generate an accurate pharmacological activity measurement. Traditionally,
the concentration of a chemical product is obtained by purifying milligram quantities to a purity greater than 95%, drying the purified material,
and weighing it as accurately as possible. A solution of known concentration is then prepared by adding a defined-volume solvent to the known
mass of the compound. This method is too restrictive and slow to keep
pace with the perpetual acceleration of the drug discovery process and is
inaccurate in the presence of impurities, counter-ions, unevaporated solvents, and unidentified particulate—particularly when compounds are
available only in minute amounts, as they are in combinatorial chemistry.
An alternative quantitation method that is sensitive, specific, and suitable
for high-throughput analysis is needed.

Application of a nonselective detector to the chromatographic analysis
of compound libraries, in theory, would improve the speed and accuracy
with which these analyses are performed, by allowing the quantitation of
the entire library of compounds with a single external standard. While no
truly universal detection method for HPLC exists, there are several detection methods that have been demonstrated to provide a similar response

to compounds within a given class. Two of these methods are discussed below.

6.4.1   HPLC with a chemiluminescence nitrogen detector (CLND)

Reversed-phase, high-performance liquid chromatography (RP-HPLC) with UV detection provides a rapid reproducible, and sensitive means of assessing the presence and sometimes the purity of compounds. The use of MS detection with RP-HPLC adds compound identity and structure verification to the analysis. However, it is difficult and cumbersome to use either of these detectors for online compound quantitation because of the requirement for a proper calibration standard, which is either the authentic compound or a closely related analogue.

HPLC with chemiluminescent nitrogen detection (HPLC-CLND) was used for the quantitative analysis of peptides and organic compounds synthesized by solid-phase peptide synthesis (Bizanek et al. 1996; Fitch et al. 1997; Taylor et al. 1998). CLND provided quantitative information regarding the nitrogen distribution of peptide samples following HPLC separation. This technique permits the universal quantitation of the nitrogen-containing compounds in an online mode without pre- or postcolumn derivatization and free of interference from nonnitrogen-containing UV chromophores. Approximately 90% of the > 65,000 developmental and marketed drugs in the commercial database MDL Drug Data Report (MDDR) contain nitrogen, indicating that the CLND is widely applicable as a detector for analyzing pharmaceuticals.

The nitrogen detection mechanism, which is based on a chemiluminescence reaction, is elucidated in Scheme 6.1. When each analyte reaches the nitrogen detector, it is mixed with oxygen and undergoes oxidation at high temperature (1000–1100°C) in the furnace. Nitrogen-containing components, with the exception of diatomic nitrogen (N_2), are converted to nitric oxide (NO). The NO is dried and reacted with ozone (O_3) in the reaction chamber to form nitrogen dioxide in the excited state ($^*NO_2$), which rapidly decays to the stable ground state (NO_2) and releases a photon (hυ). The emitted photons are detected by a photomultiplier tube (PMT), and the amplified signal is recorded using an integrator.

$$R_3N \ + \ O_2 \ \longrightarrow \ H_2O \ + \ \cdot NO \ + \ \text{other oxides}$$

$$\cdot NO \ + \ O_3 \ \longrightarrow \ \overset{*}{N}O_2 \ \longrightarrow \ NO_2 \ + \ h\upsilon$$

Scheme 6.1

By flow injection analysis (FLA) and in conjunction with reversed-phase (RP) chromatography using gradient elution, the CLND produced a linear response from 25 to 6400 pmol of nitrogen, which was equivalent for a set of chemically and structurally diverse compounds. An average error of approximately ±10% was determined for the compounds. In addition, the response was independent of mobile-phase composition.

CLND detection for HPLC is an important new high-throughput technique for measuring yields and purities in organic chemistry and combinatorial chemistry. This destructive detector can be calibrated with any nitrogen-containing compound and, thus, allows quantitative analysis when authentic reference standards are not available. The applications to library synthesis (Fitch et al. 1997; Shah et al. 2000), a mixture of 16 crude synthetic peptides produced by SPPS (Fujinari et al. 1996) and peptides that were isolated from casin hydrolysate (Fujinari and Manes 1994) were reported.

The measurement of yield is the premier application of HPLC/CLND quantitative analysis, but other applications are clear. In the early stages of a lead development project, weighable quantities of authentic pure samples of a compound are not available, and yet quantitative measurements such as IC_{50}, solubility, and plasma stability need to be made. HPLC/CLND can be used to calibrate solutions made from submilligram synthetic samples. In the split-pool method of combinatorial synthesis, mixtures of compounds are made that are difficult to characterize. The HPLC/CLND of a nominally equimolar pool (based on nitrogen) should yield equal-sized peaks. The CLND will be especially useful in conjunction with splitting to a mass spectrometer.

Ultrafast high-resolving gradient RP-HPLC with analysis times of less than 1 min has been demonstrated to be applied to the QC of combinatorial libraries. This separation method increases the throughput for purity determination by at least a factor of five compared to current technology, while maintaining sufficient resolving power to analyze a large variety of combinatorial mixture (Goetzinger and Kyranos 1998).

The precision of the CLND response when used with gradient elution RP-HPLC was measured by quadruplicate injections of a mixture containing the nine calibration compounds at equal nitrogen concentration (Taylor et al. 1998). Under the chromatographic conditions used, the compounds were all baseline-resolved and exhibited acceptable peaks shape in the water/MeOH/IPA mobile phase (Figure 6.8). Peak widths in the CLND trace were nearly identical to those in the UV trace. The baseline remained flat over the entire gradient, making quantitation relatively simple. For each compound, peak areas were measured and averaged. All compounds exhibited RSDs of ≤ 2% for the quadruplicate injections. In addition, peak areas in the CLND trace showed less variability than those in the UV trace. In the CLND trace, peak areas exhibited an average deviation of ~9% around the mean peak area, whereas, after adjusting for the mass of each compound injected, UV peak areas deviated ~15% at 214 nm and ~120% at 270 nm. These results demonstrate good reproducibility of the CLND response for replicate injections and between different compounds.

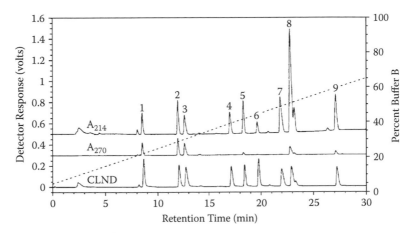

Figure 6.8 Chromatogram of a mixture of the calibration compounds separated by RP-HPLC with gradient elution and detected by UV absorption and chemi-luminescent nitrogen. The injected sample contained 2 nmol each of caffeine (1), triprolidine (2), chloropheniramine (3), diphenylalanine (4), 6-nor-6-allyllysergic acid diethylamine (5), angiotensin II (6), diphenhydramine (7), doxepin (8), and dibucaine (9). (Reproduced with permission from *Anal. Chem.* 70, 3344, 1998. Copyright© 1998, American Chemical Society.)

6.4.2   *Evaporative light scattering detector (ELSD)*

Ford and Kennard (1996) were the first to demonstrate equivalent mass response for a series of low-molecular-weight polymers using an evaporative light-scattering detector (ELSD). Briefly, the column effluent enters the detector through a narrow bore tube where it is mixed with a high-velocity stream of nitrogen gas and nebulized. The nebulized sample stream then enters a heated drift tube. The drift tube is maintained at a temperature sufficient to cause rapid evaporation of the mobile-phase solvent, leaving a narrow mist of the nonvolatile solute to be swept towards the detection region by a stream of nitrogen carrier gas. The detection system consists of a 670-nm laser and a silicon photodiode oriented at an angle of 90° and perpendicular to the central axis of the cylindrical drift tube. Passage of the narrow band of solute particulate through the laser beam results in scattering of the source. The intensity of the scattered light registered by a photodiode is proportional to the total amount of sample present in the eluting solute band.

The advantages of ELSD over UV detection for the quantitation of organic combinatorial libraries composed of small organic compounds by HPLC have been described (Kibbey 1995; Hsu et al. 1999; Cremin and Zeng 2002). The detector's response is independent of the sample chromophore, which makes it well suited to chromatographic analyses of mixtures of dissimilar solutes. Thus, HPLC with ELSD offers the potential for reducing false positive or false negative results in screening assays, because of its ability to detect the presence of impurities that absorb poorly in the UV (e.g., those impurities originating from the polymeric support). Furthermore, the ELSD exhibits a nearly equivalent response to compounds of similar structural class. Hence, rapid quantitation for compound libraries may be carried out with the use of a single external standard. The quantitation errors for amino acids by normal-phase HPLC with ELSD is in average approximately ±10%.

Successful application of the ELSD as a nonselective detector in HPLC has been used for the analysis of a wide variety of compounds, including alkyl- and alkylarylsurfactants (Bear 1988), triglycerides (Stolyhwo et al. 1985), phospholipids (Sotirhos et al. 1985), sugars (Macrae and Dick 1981), pharmaceutical salts (Peterson and Risley 1995), and excipients (Lafosse et al. 1992). In each of these applications the advantages of the ELSD have been its near-equivalent response to a variety of compounds within a given structural class, its compatibility with gradient elution, and its over-all reliability and ease of operation.

6.5   Summary

In most cases, library members should be characterized after cleavage from solid support. It is evident from the above discussions that MS is the major technique in the quality control of soluble combinatorial libraries, while HT NMR is still in its infancy. Speed and sensitivity are main advantages of the MS techniques, and quantitation is a drawback.

Both qualitative and quantitative analyses, of combinatorial library members, at least for a statistically representative set, are crucial for confirming the chemical diversity desired and ensuring confidence about screening results. Efficient quantitation methods in high-throughput format are currently not available, but quantitative detectors, such as CLND and ELSD, provide some possibilities for quantitation of combinatorial library members both in the FIA and online format.

Mixture analysis is at a very early stage of study. Comparison of mass distribution obtained from MS measurements and that from computer simulation gives a very crude approximation because the ionization efficiency of a diverse set of compounds is unknown. This opens a field for future research and development.

In general, the application of sequential analytical methods in dealing with products made from parallel synthesis is definitely a mismatch. A breakthrough in parallel qualitative and quantitative analysis is badly needed. Liquid–liquid extraction and solid-phase capture reagents are useful in special cases for small numbers of compounds. However, for libraries of a few thousand compounds, HPLC is a viable alternative. Beyond these numbers, such factors as solvent requirements, the number of fractions, and tracking become prohibitive. Supercritical fluid chromatography has been successfully employed in automated purification instrumentation and is expected to be capable of purifying libraries of tens of thousands of compounds.

References

Barnes, C.A.S., Hilderbrand, A.E., Valentine, S.J., Clemmer, D.E. 2002. Resolving isomeric peptide mixtures: A combined HPLC/Ion mobility-TOFMS analysis of a 4000-component combinatorial library. *Anal. Chem.* 74, 26.

Barnes, C.A.S., Li, J.W., Marshall, W.S., Clemmer, D.E. 2000. Determining synthetic failures in combinatorial libraries by hybrid gas-phase separation methods. *J. Am. Soc. Mass. Spectrom.* 11, 352.

Bear, G.R. 1988. Universal detection and quantitation of surfactants by high-performance liquid chromatography by means of the evaporative light-scattering detector. *J. Chromatogr.* 459, 91.

Bizanek, R., Manes, J.D., Fujinari, E.M. 1996. Chemiluminescent nitrogen detection as a new technique for purity assessment of synthetic peptides separated by reversed-phase HPLC. *Peptide Res.* 9, 40.

Cheng, X., Hochlowski, J. 2002. Current application of mass spectrometry to combinatorial chemistry. *Anal. Chem.* 74, 2679.

Chin, J., Fell, J.B., Jarosinski, M., Shapiro, M.J., Wareing, J.R. 1998. HPLC/NMR in combinatorial chemistry. *J. Org. Chem.* 63, 386.

Cremin, P.A., Zeng, L. 2002. High-throughput analysis of natural product compound libraries by parallel LC-MS evaporative light scattering detection. *Anal. Chem.* 74, 5492.

Daley, D.J., Scammell, R.D., James, D., Monks, I., Raso, R., Ashcroft, A.E., Hudson, A.J. 1997. High-throughput LC-MS verification of drug candidates from 96-well plates. *Am. Biotechnol. Lab.* 15, 24.

Demirev, P.A., Zubarev, R.A. 1997. Probing combinatorial library diversity by mass spectrometry. *Anal. Chem.* 69, 2893.

Dunayevskiy, Y., Vouros, P., Carell, T., Wintner, E.A., Rebek, J. Jr. 1995. Characterization of the complexity of small-molecule libraries by electrospray ionization mass spectrometry. *Anal. Chem.* 67, 2906.

Dunayevskiy, Y.M., Vouros, P., Wintner, E.A., Shipps, G.W., Carell, T., Rebek, J. Jr. 1996. Application of capillary electrophoresis-electrospray ionization mass spectrometry in the determination of molecular diversity. *Proc. Natl. Acad. Sci. U.S.A.* 93, 6152.

Fang, A.S., Vouros, P., Stacey, C.C., Kruppa, G.H., Laukien, F.H., Wintner, E.A., Carell, T., Rebek, J.Jr. 1998. Rapid characterization of combinatorial libraries using electrospray ionization Fourier transform ion cyclotron resonance mass spectrometry. *Combi. Chem. and High Throughput Screening* 1, 23.

Fitch, W.L. 1999. Analytical methods for quality control of combinatorial libraries. *Mol. Diversity* 4, 39.

Fitch, W.L., Szardenings, A.K., Fujinari, E.M. 1997. Chemiluminescent nitrogen detection for HPLC: An important new tool in organic analytical chemistry. *Tetra. Lett.* 38, 1689.

Franz, A.H., Liu, R.W., Song, A.M., Lam, K.S., Lebrilla, C.B. 2003. High-throughput one-bead-one-compound approach to peptide-encoded combinatorial libraries: MALDI-MS analysis of single tentagel beads. *J. Comb. Chem.* 5, 125.

Ford, D.L., Kennard, W. 1996. Vaporization analyzer, *J. Oil Color Chem. Assoc.* 49, 299.

Fujinari, E.M., Manes, J.D. 1994. Nitrogen-specific detection of peptides in liquid chromatography with a chemiluminescent nitrogen detector. *J. Chromatogr. A* 676, 113.

Fujinari, E.M., Manes, J.D., Bizanek, R. 1996. Peptide content determination of crude synthetic peptides by reversed-phase liquid chromatography and nitrogen-specific detection with a chemiluminescent nitrogen detector. *J. Chromatogr. A* 743, 85.

Furka, A., Sebestyen, F., Asgedom, M., Dibo, G. 1988. In *Highlights of Modern Biochemistry: Proceedings of the 14th International Congress of Biochemistry* (VSP, Utrecht, the Netherlands), 5, 47.

Furka, A., Sebestyen, F., Asgedom, M., Dibo, G. 1991. General method for rapid synthesis of multicomponent peptide mixtures. *Int. J. Pept. Proteins Res.* 37, 487.

Goetzinger, W.K., Kyranos, J.N. 1998. Fast gradient RP-HPLC for high-throughput quality control analysis of spatially addressable combinatorial libraries. *Am. Lab.* 30, 27.

Gorlach, E., Richmond, R., Lewis, I. 1998. High-throughput flow injection analysis mass spectroscopy with networked delivery of color-rendered results. 2. Three-dimensional spectral mapping of 96-well combinatorial chemistry racks, *Anal. Chem.* 70, 3227.

Haap, W.J., Metzger, J.W., Kempter, C., Jung, G. 1997. Composition and purity of combinatorial aryl ether collections analyzed by electrospray mass spectrometry. *Mol. Diversity* 3, 29.

Hegy, G., Gorlach, E., Richmond, R., Bitsch, F. 1996. High throughput electrospray mass spectrometry of combinatorial chemistry racks with automated contamination surveillance and results reporting. *Rapid Commun. Mass Spectr.* 10, 1894.

Hilaire, P.M.St., Cipolla, L., Tedebark, U., Meldal, M. 1998. Analysis of organic reactions by thin layer chromatography combined with matrix-assisted laser desorption/ionization time-of-flight mass spectrometry. *Rapid Commun. Mass Spectrom.* 12, 1475.

Houghten, R.A., Pinilla, C., Blondelle, S.E., Appel, J.R., Dooley, C.T., Cuervo, J.H. 1991. Generation and use of synthetic peptide combinatorial libraries for basic research and drug discovery. *Nature* 354, 84.

Hsieh, Y.F., Gordon, N., Regnier, F., Afeyan, N., Martin, S.A., Vella, G.J. 1996. Multidimensional chromatography coupled with mass spectrometry for target-based screening. *Mol. Diversity* 2, 189.

Hsu, B.H., Orton, E., Tang, S.-Y., Carlton, R.A. 1999. Application of evaporative light scattering detection to the characterization of combinatorial and parallel synthesis libraries for pharmaceutical drug discovery. *J. Chromatogr. B* 725, 103.

Kassel, D.B. 2001. Combinatorial chemistry and mass spectrometry in the 21st century drug discovery laboratory. *Chem. Rev.* 101, 255.

Keifer, P.A. 1997. High-resolution NMR techniques for solid-phase synthesis and combinatorial chemistry. *Drug Disc. Today* 2, 468.

Keifer, P.A. 1998. New methods for obtaining high-resolution NMR spectra of solid-phase synthesis resins, natural products, and solution-state combinatorial chemistry libraries. *Drugs of the Future* 23, 301.

Kenseth, J.R., Coldiron, S.J. 2004. High-throughput characterization and quality control of small-molecule combinatorial libraries. *Curr. Opin. Chem. Biol.* 8, 418.

Kibbey, C.E. 1995. Quantitation of combinatorial libraries of small organic molecules by normal-phase HPLC with evaporative light-scattering detection. *Mol. Diversity* 1, 247.

Kiplinger, J.P., Cole, R.O., Robinson, S., Roskamp, E.J., Ware, R.S., O'Connell, H.J., Brailsford, A., Batt, J. 1998. Structure-controlled automated purification of paralleled synthesis products in drug discovery. *Rapid Commun. Mass Spectr.* 12, 658.

Lafosse, M., Elfakir, C., Morin-Allory, L., Dreux, M. 1992. The advantages of evaporative light-scattering detection in pharmaceutical analysis by high-performance liquid chromatography and supercritical fluid chromatography. *J. High-Res. Chromatogr.* 15, 312.

Lam, K.S., Salmon, S.E., Hersh, E.M., Hruby, V.J., Kazmierski, W.M., Knapp, R.J. 1991. A new type of synthetic peptide library for identifying ligand-binding activity. *Nature* 354, 82.

Lambert, P.H., Bertin, S., Fauchere, J.L., Volland, J.P. 2001. Quality assessment of combinatorial pooled libraries using mass spectrometry. *Combi. Chem. and High Through. Screen.* 4, 317.

Lambert, P.H., Boutin, J.A., Bertin, S., Fauchere, J.-L., Volland, J.-P. 1997. Evaluation of high performance liquid chromatography/Electrospay mass spectrometry with selected ion monitoring for the analysis of large synthetic combinatorial peptide libraries. *Rapid Commun. Mass Spectr.* 11, 1971.

Letot, E., Koch, G., Falchetto, R., Bovermann, G., Oberer, L., Roth, H.J. 2005. Quality control in combinatorial chemistry: Determinations of amounts and comparison of the "purity" of LC-MS-purified samples by NMR, LC-UV and CLND. *J. Comb. Chem.* 7, 364.

Lewis, K.C., Fitch, W.L., Maclean, D. 1998. Characterization of a split-pool combinatorial library. *LC-GC* 16, 644.

Lin, M., Shapiro, M.J. 1997. Mixture analysis by NMR spectroscopy. *Anal. Chem.* 69, 4731.

Lin, M., Shapiro, M.J. 1996. Mixture analysis in combinatorial chemistry: Application of diffusion-resolved NMR spectroscopy. *J. Org. Chem.* 61, 7617.

Lindon, J.C., Farrant, R.D., Sanderson, P.N. 1995. Separation and characterization of compounds of peptide libraries using on-flow coupled HPLC-NMR spectroscopy. *Magn. Reson. in Chem.* 33, 857.

Luo, H., Chen, L., Li, Z.Q., Ding, Z.S., Xu, X.J. 2003. Frontal immunoaffinity chromatography with mass spectrometric detection: A method for finding active compounds from traditional Chinese herbs. *Anal. Chem.* 75, 3994.

Macrae, R., Dick, J. 1981. Analysis of carbohydrates using the mass detector. *J. Chromatogr.* 210, 138.

Mathur, S., Hassel, M., Steiner, F., Hollemeyer, K., Hartmann, R.W. 2003. Development of a new approach for screening combinatorial libraries using MALDI-TOF-MS and HPLC-ESI-MS/MS. *J. Biomol. Screen.* 8, 136.

Metzger, J.W., Kempter, C., Wiesmuller, K.-H., Jung, G. 1994. Electrospray mass spectrometry and tandem mass spectrometry of synthetic multicomponent peptide mixtures: Determination of composition and purity. *Anal. Biochem.* 219, 261.

Metzger, J.W., Wiesmuller, K.-H., Gnau, V., Brunjes, J., Jung, G. 1993. Ion-spry mass spectrometry and high-performance liquid chromatography-mass spectrometry of synthetic peptide libraries. *Angew. Chem. Chem. Int. Ed. Engl.* 32, 894.

Nawrocki, J.P., Wigger, M., Watson, C.H., Hayes, T.W., Senko, M.W., Benner, S.A., Eyler, J.R. 1996. Analysis of combinatorial libraries using electrospray Fourier transform ion cyclotron resonance mass spectrometry. *Rapid Commun. Mass Spectr.* 10, 1860.

Peterson, J.A., Risley, D.S. 1995. Validation of an HPLC method for the determination of sodium in LY293111 sodium, a novel LTB$_4$ receptor antagonist using evaporative light-scattering detection. *J. Liq. Chromatogr.* 18, 331.

Robinson, G.M., Manica, D.P., Taylor, E.W., Smyth, M.R., Lunte, C.E. 1998. Development of a capillary electrophoretic separation of an N-(substituted)-glycine-peptoid combinatorial mixture. *J. Chromatogr. B* 707, 247.

Schriemer, D.C., Bundle, D.R., Li, L., Hindsgaul, O. 1998. Micro-scale frontal affinity chromatography with mass spectrometric detection: A new method for the screening of compound libraries. *Angew. Chem. Int. Ed.* 37, 3383.

Shah, N., Gao, M., Tsutsui, K., Lu, A., Davis, J., Scheuerman, R., Fitch, W.L. 2000. A novel approach to high-throughput quality control of parallel synthesis libraries. *J. Comb. Chem.* 2, 453.

Shin, Y.G., Breemen, R.B. 2001. Analysis and screening of combinatorial libraries using mass spectrometry. *Biopharm. Drug Dispos.* 22, 353.

Simms, P.J., Jeffries, C.T., Huang, Y.J., Zhang, L., Arrhenius, T., Nadzan, A.L. 2001. Analysis of combinatorial chemistry samples by micellar electrokinetic chromatography. *J. Comb. Chem.* 3, 427.

Smallcombe, S.H., Patt, S.L., Keifer, P.A. 1995. Wet solvent suppression and its applications to LC NMR and high-resolution NMR spectroscopy. *J. Magn. Reson. A.* 117, 295.

Smart, S.S., Mason, T.J., Bennell, P.S., Maeij, N.J., Geysen, H.M. 1996. High-throughput purity estimation and characterization of synthetic peptides by electrospray mass spectrometry. *Int. J. Peptide Res.* 47, 47.

Sotirhos, N., Thorngren, C., Herslof, B. 1985. Reversed-phase high-performance liquid chromatographic separation and mass detection of individual phospholipid classes. *J. Chromatogr.* 331, 313.

Stolyhwo, A., Colin, H., Guiochon, G. 1985. Analysis of triglycerides in oils and fats by liquid chromatography with the laser light-scattering detector. *Anal. Chem.* 57, 1342.

Tang, L., Fitch, W.L., Smith, P., Tumelty, D., Cao, K., Ferla, S.W. 2001. On-line sample preparation for high throughput reversed-phase lc/ms analysis of combinatorial chemistry libraries. *Comb. Chem. and High Through. Screen.* 4, 287.

Taylor, E.W., Qian, M.G., Dollinger, G.D. 1998. Simultaneous on-line characterization of small organic molecules derived from combinatorial libraries for identity, quantity, and purity by reversed-phase HPLC with chemiluminescent nitrogen, UV, and mass spectrometric detection. *Anal. Chem.* 70, 3339.

Triolo, A., Altamura, M., Cardinali, F., Sisto, A., Maggi, C.A. 2001. Mass spectrometry and combinatorial chemistry: A short outline. *J. Mass Spectrom.* 36, 1249.

Tutko, D.C., Henry, K.D., Winger, B.E., Stout, H., Hemling, M. 1998. Sequential mass spectrometry and MS^n analyses of combinatorial libraries by using automated matrix-assisted laser desorption/ionization Fourier transform mass spectrometry. *Rapid Commun. Mass Spectr.* 12, 33.

Wang, T., Zeng, L., Strader, T., Burton, L., Kassel, D.B. 1998. A New ultra-high throughput method for characterizing combinatorial libraries incorporating a multiple probe autosampler coupled with flow injection mass spectrometry analysis. *Rapid Commun. Mass Spectr.* 12, 1123.

Weller, H.N., Young, M.G., Michalczyk, S.J., Reitnauer, G.H., Cooley, R.S., Rahn, P.C., Loyd, D.J., Fiore, D., Fischman, S.J. 1997. High throughput analysis and purification in support of automated parallel synthesis. *Mol. Diversity* 3, 61.

Winger, B.E., Campana, J.E. 1996. Characterization of combinatorial peptide libraries by electrospray ionization Fourier transform mass spectrometry. *Rapid Commun. Mass Spectrom.* 10, 1811.

Wu, Q. 1998. Multistage accurate mass spectrometry: A "basket in a basket" approach for structure elucidation and its application to a compound from combinatorial synthesis. *Anal. Chem.* 70, 865.

Xue, G., Pang, H.M., Yeung, E.S. 1999. Multiplex capillary zone electrophoresis and micellar electrokinetic chromatography with internal standardization. *Anal. Chem.* 71, 2642.

Xu, J.Y., Braun, R.M., Winograd, N. 2003. Applicability of imaging time-of flight secondary ion MS to the characterization of solid-phase synthesized combinatorial libraries. *Anal. Chem.* 75, 6155.

Yan, B., Fang, L.L., Irving, M., Zhang, S., Boldi, A.M., Woolard, F., Johnson, C.R., Kshirsagar, T., Figliozzi, G.M., Krueger, C.A., Collins, N. 2003. Quality control in combinatorial chemistry: determination of the quantity, purity, and quantitative purity of compounds in combinatorial libraries. *J. Comb. Chem.* 5, 547.

Yates, M., Wislocki, D., Roberts, A., Berk, S., Klatt, T., Shen, D.M., Willoughby, C., Rosauer, K., Chapman, K., Griffin, P. 2001. Mass spectrometry screening of combinatorial mixtures, correlation of measured and predicted electrospray ionization spectra. *Anal. Chem.* 73, 2941.

Zeng, L., Kassel, D.B. 1998. Developments of a fully automated parallel HPLC/ mass spectrometry system for the analytical characterization and preparative purification of combinatorial libraries. *Anal. Chem.* 70, 4380.

chapter seven

High-throughput purification

7.1 Introduction

One of the driving forces in applying combinatorial chemistry to drug discovery is to accelerate lead discovery and preclinical research. The synthesis of combinatorial libraries by parallel synthesis, followed by high-throughput biological screening, has become a paradigm for early drug discovery. It is important that these combinatorial library compounds are pure in order to obtain high quality assay data. Any impurity in samples may lead to false positive results and wrong IC_{50} values. This may also lead to incorrect structure-activity relationship (SAR) conclusions. Even with advances in solid-phase (Thompson et al. 2000) and solution-phase (Baldino et al. 2000) synthesis methods and intensive reaction optimization (Yan et al. 2001), excess reagents, starting materials, synthetic intermediates, and byproducts are often found with the desired product. Furthermore, strong solvents for swelling the resin beads or scavenging treatment in solution-phase reactions can often bring in additional impurities extracted from resins and plastic plates. The requirement for *the absolute purity* (Yan et al. 2003) of combinatorial library compounds demands the development of high-throughput purification (HTP) methods (Weller et al. 1999) at a scale that matches combinatorial or parallel synthesis (10–50 mg). An HTP method for purifying combinatorial libraries must be generally applicable for all chemistries and structural types, rapid, easily automated, and capable of yielding pure compounds in a single pass at low cost (Yan et al. 2004). In this chapter, methods, strategies, and recent approaches to HTP are reviewed.

7.2 Nonchromatographic high-throughput purification methods

For compounds synthesized using conventional methods, organic/aqueous extraction is often used to remove excess reagents. Extraction methods (liquid–liquid extraction and solid supported liquid–liquid extraction) (Booth et al. 1999; Parlow et al. 1999; Thompson et al. 2000), scavenger methods (Thompson et al. 2000), solid–liquid phase extraction (Thompson et al. 2000), and solid-phase extraction (Nilsson et al. 2000) have also been employed for the purification of combinatorial libraries (Weller et al. 1999; Hughes et al. 2001).

Solid-supported liquid–liquid extraction is a useful method for removing impurities. Organic acid and organic base contaminant can be effectively removed by this technique, which involves supporting an aqueous buffer stationary phase absorbed on a diatomaceous earth solid support, and subsequent application of the reaction mixture followed by organic mobile phase elution. This SLE technique was first reported by Johnson et al. (1998) in an automated high-throughput format, for the purification of a library of N-alkylaminoheterocycles.

An automated high-throughput liquid–liquid extraction methodology has been developed and utilized for the removal of excess amines and water-soluble impurities from combinatorial library samples (Peng et al. 2000). After evaluation of different extraction solvents and devices, optimal extraction conditions were obtained by employing butyl acetate as an organic solvent and hydrochloric acid as an aqueous solvent in a diatomaceous-earth-packed 96-well plate with a bottom hydrophobic filter. This automated method was applied to the initial purification of crude combinatorial library samples in a 96-well format. Among nine amines tested, greater than 98% removal of excess amines was achieved and more than 85% recovery and better than 85% purity (average purity of 90%) were obtained from the partial combinatorial library samples. The methodology can be modified to remove acidic and neutral water-soluble components by selecting appropriate aqueous extraction solvents. In conjunction with parallel preparative HPLC techniques for further hit purification and identification, this automated LLE methodology should provide great benefits to initial purification of crude library samples for high-throughput screening, which would ultimately result in reduced labor, time, and cost associated with the purification of combinatorial libraries.

Conventional LLE requires precise removal of the aqueous layer, which is not amenable to a large number of samples. One distinctive advantage for SLE over the more traditional LLE is its ease of automation and parallel processing. A parallel format SLE technique has been employed for the rapid purification of compound libraries (Breitenbucher et al. 2001).

Solid-phase scavenger methods have been employed with increasing frequency as a preliminary reaction cleanup step in combinatorial chemistry (Deal et al. 1998; Breitenbucher et al. 2001; Mukherjee et al. 2003; Pirrung et al. 2003), and have become commercially available (Argonaut, Calbiochem-Novabiochem, Varian, Alltech). Kaldor et al. (1996) first reported on this approach, employing solid supported electrophiles and nucleophiles for reaction purification in acylation and alkylation reactions. Yield and purity values reported were 90–95% and 50–99%, respectively, for a library generated by reductive amination. Booth et al. (1997) achieved the removal of known reaction product impurities by the application of custom-synthesized polymer supported reagents, specially polystyrene-divinylbenzene-supported derivatives of methylisocyanate and tris(2-aminomethyl)amine for cleanup of byproducts resulting from urea, thiourea, sulfonmide, amide, and pyrazole libraries.

LLE and SLE also suffer from some limitations. There are often hydrophobic byproducts, such as that from an incomplete removal of a protecting group in combinatorial samples. These impurities will not be removed by SLE and will affect the product purity. On the other hand, hydrophilic samples with low logPs may get lost during the process (Yan et al. 2003).

Solid-phase extraction (SPE) is a technique similar to solid-phase scavenger cleanup, but differs in that impurities removed are not covalently bound to the resin, as is most frequently the case with scavenger techniques. SPE packing materials are commercially available or easily set up in various formats, including cartridges and microtiter plates, each of which is amenable to facile automation. Researchers at Metabasis have reported (Bookser et al. 2001) a detailed SPE investigation into the applicability of 11 ion exchange resins for the efficacy of recovering carboxylic acids from parallel format syntheses. A microtiter well batch-wise format was employed throughout, and demonstrated achievable purity levels of up to 98% by this capture-release process, with recovery typically on the order of 50–80%.

The above-mentioned techniques all constitute methodologies that can routinely achieve only preliminary reaction cleanup or the removal of library specific components, although in certain cases they give quite good results concerning achievable yield and purity. In contrast, automated high-performance liquid chromatography (HPLC) and supercritical fluid chromatography (SFC) are more generally amenable to the purification of a wide range of structural types and routinely provide compounds of 95% or better purity.

7.3   Chromatographic high-throughput purification methods

Although methods mentioned in the previous section have been used for purification of crude reaction mixtures, chromatographic techniques typically provide pure compounds more reliably. Over the years, semipreparative reversed-phase high-performance liquid chromatography (RP-HPLC) has become the most popular technique for purifying combinatorial libraries owing to the high resolution of HPLC, the ease with which HPLC instruments can be interfaced with automatic sampling and fraction collecting devices for unattended operation, and the possibility of developing a generic method for a whole library or even many libraries (Zeng et al. 1998; Xu et al. 2002; Blom et al. 2003; Schaffrath et al. 2005). The general process of chromatographic purification consists of three major parts (Figure 7.1): the separation of the sample components, an effective fraction collection strategy, and the post-purification processes (including fraction qualification, evaporation, weighing, and transfer to a proper vial format) (Karancsi et al. 2005).

High-throughput HPLC purification systems described to date all rely on a detection methodology that recognizes the elution of a peak of interest and, based on this signal, initiates the collection of eluting components (Hughes et al. 2001). Detection methods employed include ultraviolet (UV) spectroscopy, mass spectrometry (MS), and evaporative light-scattering detection (ELSD). The HPLC systems are almost universally reversed-phase systems, and the first high-throughput system was described by Weller et al. (1997). This open-access HPLC system was based on UV-triggered fraction collection, and it was capable of unattended operation. Fast flow rates, short columns, and rapid universal reversed-phase gradient elution methods allowed the purification of up to 200 samples a day. An HTP system was developed by Yan et al. (2004) to purify combinatorial libraries at a 50-/100-mg scale with a throughput of 250 samples/instrument/day using an "accelerated retention window" (ARW) method. Using a UV and ELSD-triggered fraction collection, Yan and co-workers applied the ARW method to shorten the purification time and targeted one fraction per injection to simplify data tracking, lower QC workload, and simplify the postpurification processing. The accurate retention time and peak height for all compounds were first determined using an eight-channel parallel LC/UV/MS system, and the specific preparative HPLC conditions for individual compounds were calculated. The preparative HPLC conditions include the compound-specific gradient segment for individual compounds with a fixed gradient slope and the compound-specific UV or ELSD threshold for triggering a fraction collection device. A unique solvent composition or solvent strength was programmed for each compound in the preparative HPLC in order to elute all compounds at the same target time. Considering the possible deviation of the predicted retention time, a 1-min window around the target time was set to collect peaks above a threshold based on UV or ELSD detection. More than 500,000 drug-like compounds have been purified using this system during a period of three years. In addition to the UV-triggered fraction collection method (Edwards et al. 2003; Yan et al. 2004; Karancsi et al. 2005), the increased specificity of mass-directed purification provided fractions expected to contain the desired compound on the basis of m/z. Because of the simplification of this method, the majority of operations nowadays are using this method (Zhang et al. 2006; Espada et al. 2008; Guth et al. 2008).

To develop structure activity relationships (SAR) for a compound series, each library member should preferably be >95% pure. Historically, achieving and quantifying such high-purity criteria for each library member proved to be the slow step for most lead discovery groups. To address this issue, significant modifications have been made by Leister et al. (2003) to a commercial Agilent preparative LC/MS system to allow for the general mass-guided purification of diverse compound libraries. The modifications include (1) the "DMSO slug" approach for the purification of samples with poor solubility; (2) an active splitter to reduce system back-pressure, reduce the delay volume, and allow for a variable split ratio; (3) a sample loading pump for the quick purification of dilute samples; (4) a preparative column-selection valve to quickly change column selectivity or sample loading; and (5) an analytical injector with a separate flow path for crude reaction or fraction analyses. The modified systems afforded excellent mass recoveries (70–95%) and purities (>95%) with a capacity to purify ~6 samples/hour per instrument.

The Biotage Parallex HPLC was reported as the first high-throughput purification system to be introduced, running four columns in parallel and enabling purification of up to 40 samples/hour (Schultz et al. 1998; Coffey et al. 1999). Although this system can be used for reversed-phase, normal phase, and ion exchange applications, reversed phase is usually preferred. The flow rate range of the standard configuration is 5–50 mL/min/flow stream with a pressure limit of 4000 psi, allowing use of a wide range of column sizes and packing materials. Samples are introduced from a deep-well microtiter plate into the four loops sequentially during a method cycle and injected simultaneously at the beginning of the next method cycle to maximize use of time. Performance and reproducibility of the Biotage Parallex was evaluated by purifying an array of 4320 compounds produced by an 11-step solid-phase synthesis in Irori MicroKans. Of the compounds, 93% were successfully processed, with >90% having purity >95% (Edwards et al. 2003).

A high-throughput purification platform was designed by ArQule with a process based on mass directed fractionation with a one-to-one mapping of sample injection to fraction collection (Goetzinger et al. 2004). A series of recovery studies validated the reliability of the mass-directed fractionation. Enforcement of one fraction per sample collection has greatly simplified the purification operation and accelerated the postpurification process. This approach has been used successfully to purify over half a million compounds in two years and resulted in a postpurification average purity of over 97% when assessed by HPLC and UV low-wavelength, which is currently the most stringent quality criteria applied in a high throughput mode.

An automated high-throughput purification system (Irving et al. 2004) was developed and optimized for the purification of a 4-amino-pyrrolidone library, which was intentionally synthesized as pairs of diastereomers by solution-phase parallel syntheses (Scheme 7.1). A total of 2592 4-amido-pyrrolidinones were ultimately isolated as single diastereomers from a matrix of 1920 syntheses. After the four-step synthesis and HPLC purification, the average yield of single pure diastereomers was 36.6%. The average chemical purity was >90%, and the average diastereomeric purity was >87%.

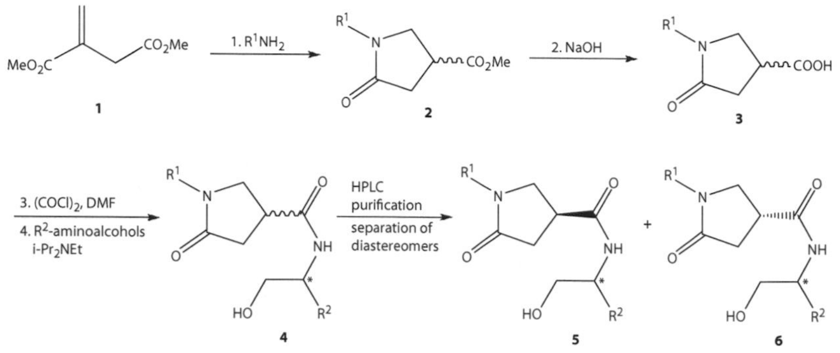

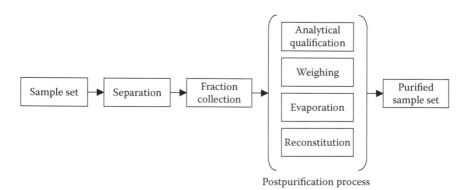

Scheme 7.1

In order to support high-throughput library purification, a novel UV-triggered fraction collection method was developed using a maximum seeking-algorithm and a six-port valve, where only the largest chromatographic peak was collected (Karancsi et al. 2005). This straightforward strategy achieves the one sample–one fraction result, thus resulting in a simpler and less error prone workup procedure. The effectiveness of this main component fraction collection method was illustrated by the results of the purification of compound libraries (altogether 6086 compounds, having an averaged rate of 79.4%).

A high-throughput system was reported for the analysis and purification of compounds. This system addressed most major bottlenecks within the process (Koppitz et al. 2005): sample tracking, data transfer, and the generation of sample/worklists using a software via barcode support. The

Figure 7.1 The general overview of the purification process.

system consisted of an analytical four-channel multiplexed electrospray (MUX) LC/MS system, a four-channel preparative LC/MS system, a liquid-handling device with attached microtiter plate (MTP) hotel for fraction pooling, and an automated solvent and waste management system. Similar to the "accelerated retention window" approach developed by Yan et al. (2004), the assignment of optimized gradients for preparative HPLC/MS runs using the analytical prescreen allows the rapid and efficient purification of compound libraries. Even peptides that often represent difficult cases for purification would be purified on our system on a 150-mg scale. Purity assessment of product fractions allowed cherrypicking of pure fraction tubes (See Figure 7.2). The pooling algorithm allowed the automated combination of pure product fractions and various scenarios for isomer pooling. Last but not least, an automated solvent and waste management system was implemented that was capable of running this system on a 24 h/day basis and contributed significantly to solving restrictions in manpower.

Although two- and four-channel LC/MS systems had the capability of purifying more compounds, it was found that doubling the number of purification channels did not always double the throughput. As the purification throughput increased, postpurification sample handling was found to be the rate-limiting step (Isbell et al. 2002). A reliable, integrated, high-throughput purification process was developed that could keep pace with the throughput achievable with the "purification factory": a four-channel LC/MUX-MS purification system, as shown in Figure 7.3 (Isbell et al. 2005). Although a peak daily throughput of 704 samples was achieved with this system, purification and processing of 528 samples/day was more realistic. The four-channel LC/MS purification system is capable of providing results comparable to those of four single-channel systems. The high-speed tube-sorting robot provided the opportunity to maximize the impact of sample purification by limiting purification to those compounds for which it would have the greatest impact. It is also an important tool for rapidly preparing high-throughput screening plates from the purified samples because it is capable of dissolving compounds and making an equal solution on the basis of the number of micromoles.

7.4   Supercritical fluid chromatographic purification methods

Along with the mainstream development of RP-HPLC-based purification, there has been an increasing interest in exploring the potential of using supercritical fluid chromatography (SFC) as an alternative to HPLC in the analysis and purification of combinatorial chemistry libraries. In contrast to reversed-phase HPLC, SFC is a normal phase chromatography technique using a compressible fluid as the mobile phase. Typically, carbon

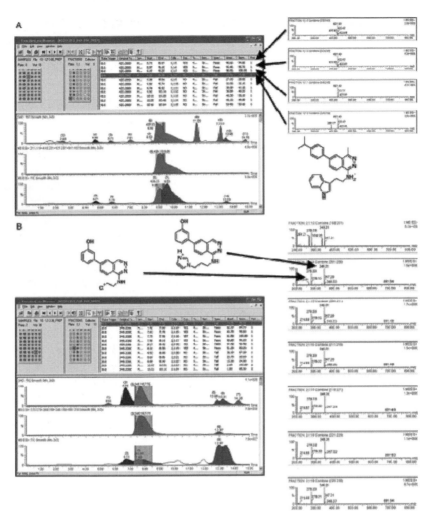

Figure 7.2 Purity assessment of fractions by postrun analysis of prep. LC/MS data. An example for a stable (A) and an easily fragmenting compound (B) is given. (Reproduced with permission from *J. Comb. Chem.* 2005, 7, 714–720. Copyright© 2005 American Chemical Society.)

dioxide (CO_2) is used as one component of the mobile phase, and methanol or methanol with an organic modifier is used as the other component. Supercritical fluids have intermolecular forces lower than those in normal liquids, resulting in lower viscosities and higher diffusion rates for the mobile phase used in SFC compared to HPLC. SFC systems can therefore be run at higher flow rates than an equivalent HPLC system without excessive column backpressure or loss of resolution, leading to shorter run times per sample. This throughput advantage has been reported in

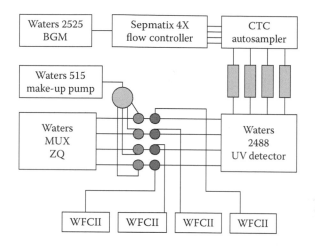

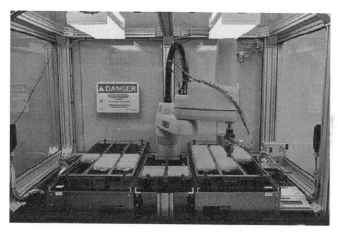

Figure 7.3 Configuration of the Purification Factory (top) and Custom GNF tube labeling, sorting, and weighing robot (bottom). (Reproduced with permission from *J. Comb. Chem.* 2005, 7, 210–217. Copyright© 2005 American Chemical Society.)

the use of analytical SFC and SFC/MS for the analysis of combinatorial libraries. Preparative SFC has the further advantage over preparative HPLC of producing purified fractions in a solvent that can be more readily removed. Evaporation of CO_2 occurs upon fraction collection, leaving purified compounds as solutions in methanol, which can be dried down much more rapidly than the aqueous fractions resulting from RP-HPLC. In addition, SFC produces significantly lower amounts of hazardous waste solvent for disposal than does HPLC, which becomes a significant advantage as compound numbers increase (Searle et al. 2004; Li and Hsieh. 2008; Taylor 2008).

Ever since Berger Instruments (Newark, DE) developed the first semi-preparative scale high-throughput SFC instrument for the purification of combinatorial chemistry libraries (Berger et al. 2000), several groups have reported the use of preparative SFC for the purification of compound libraries with systems ranging from those using simple UV-triggered fractionation to mass-directed fraction collection and parallel sample processing. SFC began to show promise as a good technology for the purification of combinatorial chemistry (Farrell et al. 2001; Hochlowski et al. 2003; Olson et al. 2002; Wang et al. 2001). The commercially available system developed by Berger Instruments is composed of several modified components to achieve chromatography where a compressible fluid is the primary mobile phase. The system comprises a pump to achieve accurate delivery of carbon dioxide, a high-pressure mixing column, a manual injector, and a UV detector with a high-pressure flow cell. Back-pressure regulation is employed to control the column outlet pressure downstream of the UV detector, and a Berger Instrument's "separator" prevents aerosol formation as fluid-phase methanol/carbon dioxide expands to the gas phase and separates from the liquid organic modifier. Pseudo-universal gradients and automatic peak detection allowed fraction collection without a scouting chromatogram or operator intervention. Peak purity exceeding 99%, and a collection efficiency of at least 95% was shown to be possible (Berger et al. 2000). Fractions are collected at an elevated pressure into a "cassette" system made up of four individual compartments, each with a glass collection tube insert, allowing efficient collection of up to four components per chromatogram.

Using such SFC systems, combinatorial libraries constructed by parallel synthesis have been successfully purified (Farrell et al. 2001). Prepurification analytical SFC/MS/CLND allows the triage of samples for purification, and an in-house software package analyzes data for predicted quality based on an evaluation of UV and MS data for the potential of co-eluting peaks during purification. This same software package selects a collection time window for purification, which is necessary to limit the number of fractions per sample. This system accommodates the purification of samples. Postpurification analytical SFC/MS/CLND is used as well to validate purified samples.

Wang et al. (2001) developed an automated packed-column semi-preparative SFC/MS system incorporating mass-directed fraction collection for the high-throughput purification of compound libraries. Incorporating a single quadrupole mass spectrometer and SFC, separations were achieved using a binary solvent system consisting of carbon dioxide and methanol.

A diverse set of 16 high-throughput organic synthesis libraries, consisting of 48 samples per library on a scale of 16–21 mg, has been purified by both preparative SFC and preparative HPLC (Searle et al. 2004). The relative effectiveness of these two purification techniques was evaluated in terms of success, yield, and purity of final product. SFC purification was carried out using a modified Berger Instruments PrepSFC system (Olson et al. 2002). A manual version of the Berger system was integrated with a Gilson 232 Autosampler for sample injection and a Cavro MiniPrep pipettor customized for fraction collection at atmospheric pressure. For each library, the number of samples in which the technique was successful, that is, succeeded in isolating the desired product, and the average yield for purified products are summarized in Table 7.1. In general, both HPLC and SFC were successful in obtaining purified products, and no single technique offers a clear advantage chromatographically. The average purified yield for the 16 libraries was virtually identical for both techniques: 34% for HPLC and 35% for SFC. On the basis of the results, a decision has been made at Abbott Laboratories to increase capacity for SFC purification; this was also due to other advantages inherent in the technology: specifically, the ease with which solvent can be evaporated from fractions and the ability to provide products in a salt-free form. The latter advantage may be important for compound storage where TFA salts are undesirable for biological screening and storage.

Table 7.1 Summary of Library Purification Results by Preparative HPLC and Preparative SFC (Searle et al. 2004)

Library No.	Sample Purified	Preparative HPLC		Preparative SFC	
		Success	Yield (%)	Success	Yield (%)
1	45	42	18	44	34
2	48	45	49	45	38
3	48	43	44	37	39
4	48	48	56	48	64
5	45	44	31	44	45
6	48	41	22	44	26
7	44	40	25	40	27
8	47	38	30	43	32
9	47	42	21	45	13
10	48	31	29	29	27
11	48	48	30	45	25
12	36	36	63	35	53
13	48	44	43	41	34
14	48	46	24	9	
15	48	45		46	
16	48	44	24	18	

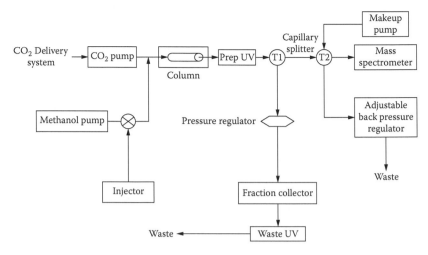

Figure 7.4 SFC/MS purification system diagram. Tubing internal diameters: auto sampler to column, column to prep UV, 0.01 in.; prep UV to T1, 0.02 in.; T2 to mass spectrometer: 0.0025 in.; T1 to pressure regulator, 0.03 in.; pressure regulator to fraction collector, 0.045 in.; fraction collector to waste UV, 0.08 in. (Reproduced with permission from *J. Comb. Chem.* 2006, 8, 705–714. Copyright© 2006 American Chemical Society.)

Zhang et al. (2006) have also developed a preparative SFC/MS system that allows real-time fractionation based on an MS signal. Issues regarding the software integration and the SFC/MS interface (Figure 7.4) have been resolved. A second UV detector, located after the fraction collector, was used in assessing the delay time and monitoring collection in real time. Column efficiency was optimized through flow path modification. The use of a wide-bore collection needle was found to significantly reduce the linear flow rate at the collector and minimize aerosol formation. Overall recoveries were >85% at flow rates up to 30 mL/min. Generally good peak shape and column loadability were achieved with 2-ethylpyridine columns without any mobile phase additive.

7.5   *Purification methods based on fluorous chemistry separation technique*

Fluorous chemistry as a new solution-phase technology has been integrated to many research areas such as the recovery of catalysts, purification of small molecules, immobilization of biomolecules, and development of new microfluidic devices and nanomaterials (Zhang et al. 2007). The development of fluorous technologies for high-throughput organic synthesis was pioneered by Curran (Curran et al. 1996; Studer et al. 1997) at the University of Pittsburgh from the mid-1990s and has been continuously developing at Fluorous Technologies, Inc. (FTI) since 2000. As an implementation of traditional solution-phase reactions in combinatorial chemistry library synthesis, the success of fluorous synthesis depends largely on the efficiency of fluorous separations. Fluorous separations are based on fluorine–fluorine interactions between the fluorous molecules and the fluorous separation media. The separation media can be fluorous solvents used for fluorous liquid–liquid extraction or fluorous silica gel used for solid-phase extraction, flash chromatography, and HPLC. Early fluorous synthesis technologies relied on "heavy fluorous" tags (60% or more fluorine by molecular weight) and liquid–liquid extraction with fluorous solvents to separate fluorous molecules from non-fluorous organics (Cavazzini et al. 1999; Curran et al. 1998). Later, "light fluorous" synthesis (40% or less fluorine by molecular weight), shorter fluorous chains (C_3F_7 to $C_{10}F_{21}$) were introduced. Liquid–liquid extraction was replaced by fluorous silica gel-based solid-phase extraction or HPLC (Curran et al. 2001; Tzschucke et al. 2002). For example, FluoroFlash silica gel has a bonded phase of C_8F_{17} perfluorohydrocarbon chains that has the capability to retain the fluorous molecules by strong fluorine–fluorine interactions. Since no fluorous solvents are required for reactions and separations, the "light fluorous" method has quickly gained popularity in parallel and combinatorial synthesis (Zhang et al. 2003).

7.5.1 Fluorous liquid–liquid extraction (F-LLE)

Fluorous media are orthogonal to organic and aqueous phases. Many fluorous solvents exhibit temperature-dependent miscibility with organic solvents. This characteristic has been utilized in designing fluorous biphasic catalysis. The separation of the reaction mixture is conducted with an organic/fluorous biphasic extraction or an organic/aqueous/fluorous triphasic extraction if water-soluble materials are involved. Fluorous liquid–liquid extractions are targeted towards heavier fluorous molecules, which have a good partition coefficient between fluorous and organic solvents. Commonly used fluorous solvents are perfluoroalkanes, perfluoroethers, and perfluoroamines such as perfluorohexane (FC-72), perfluoroheptane, and perfluoromethylcyclohexane (PFMC) (Zhang et al. 2004).

7.5.2 Fluorous solid-phase extraction (F-SPE)

F-SPE separations can be grouped into two classes: standard and reverse (Zhang et al. 2006). The standard or original solid-phase extraction involves the partitioning between a fluorous solid phase and a fluorophobic liquid phase. F-SPE was introduced in 1997 and involves the use of a fluorous solid phase and a fluorophilic (but not fluorous) solvent.

The fluorous solid phase is typically silica gel with a fluorocarbon bonded phase ($-Si(CH_3)_2(CH_2)_2C_8F_{17}$), and this is commercially available from Fluorous Technologies, Inc. under the trade name of FluoroFlash®. FluoroFlash silica gel has a bonded phase of $Si(CH_3)_2CH_2CH_2C_8F_{17}$. A reaction mixture containing fluorous molecules bearing tags such as C_6F_{13} or C_8F_{17} can be easily separated from the non-fluorous molecules by F-SPE. In a typical F-SPE separation, a crude reaction mixture is loaded onto the SPE cartridge with a minimum amount of organic solvent. The cartridge is then eluted with a fluorophobic solvent such as 80:20 MeOH/H_2O for the non-fluorous compounds, followed by a more fluorophilic solvent such as MeOH, acetone, acetonitrile, or THF for fluorous compounds. The suggested loading (amount of crude product vs. silica gel) for F-SPE separation is between 5 and 10%. The cartridge can be conditioned for reuse.

A mixture containing fluorous molecules with an Rf chain, such as C_6F_{13} or C_8F_{17}, can be easily separated from non-fluorous molecules by SPE with FluoroFlash cartridges. In a typical separation, a crude reaction mixture is loaded onto a cartridge with a minimum amount of solvent (less than 20% of silica gel volume), and then the cartridge is eluted with a "fluorophobic solvent" such as 80:20 MeOH–H_2O to remove the non-fluorous compounds. Final elution with a "fluorophilic solvent" such as MeOH, acetone, acetonitrile, or THF provides the product. Depending on the chemistry, the desired product can be a non-fluorous compound in the first fraction (MeOH–H_2O) or a fluorous compound in the second fraction (MeOH). A demonstration with a mixture of non-fluorous (blue) and fluorous (orange) aminoanthraquinone dyes is schematically shown in Figure 7.5. The separation of these two dyes on a FluoroFlash cartridge is very straightforward: the non-fluorous blue dye is readily eluted with 80:20 MeOH–H_2O (Figure 7.5, left), while the fluorous dye is retained until it is washed out with 100% MeOH (Figure 7.5, middle and right). It is noteworthy that on a C18 reversed-phase HPLC column the elution order of these two dyes is reversed: the fluorous orange dye comes out first. This result demonstrates that F-SPE is a tag-based instead of a polarity-controlled separation process (Zhang 2003).

A typical loading for SPE is between 5 and 15% of the silica by weight. The cartridge can be regenerated by washing with a strong solvent, such as MeOH, THF, and acetone. F-SPE has been widely used in the parallel separation of reaction mixtures involving fluorous reagents, catalysts, scavengers, tags, and protecting groups. Plate-to-plate F-SPE can further enhance its utility in high-throughput purification.

A commercially available Argonaut VacMaster-96™ plate-to-plate SPE station equipped with twenty-four FluoroFlash® cartridges can also be employed for parallel purification of fluorous reaction mixtures (Figure 7.6; Zhang et al. 2005). Each cartridge charged with 3 g of fluorous silica gel

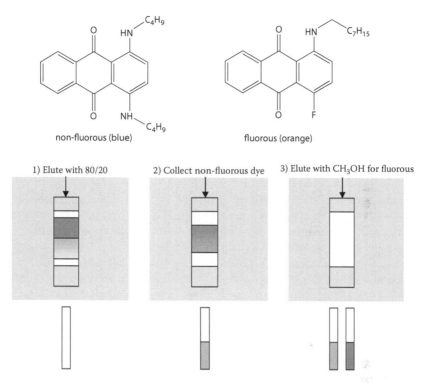

non-fluorous (blue)   fluorous (orange)

1) Elute with 80/20   2) Collect non-fluorous dye   3) Elute with CH₃OH for fluorous

Figure 7.5 Schematic diagram of F-SPE separation of non-fluorous (blue) and flu-orous (orange) dye. (1) Elute with 80/20 CH₃OH/H₂O; (2) collect non-fluorous dye; (3) elute with CH₃OH for fluorous dye.

(a) a 4 × 6 VacMaster™ system connecting   (b) 10 mL receiving well containing blue dye
to a Wob-L® piston vacuum pump   for cross-contamination test

Figure 7.6 Plate-to-plate SPE system: (A, left) a 4 × 6 VacMaster system connecting to a Wob-L piston vacuum pump and (B, right) 10-mL receiving well containing blue dye for cross-contamination test. (Reproduced with permission from *J. Comb. Chem.* 2005, 7, 893–897. Copyright© 2005 American Chemical Society.)

has the capability of producing up to 100 mg of purified small molecules. The 24-well receiving plate has a standard footprint that can be directly concentrated in a Genevac vacuum centrifuge. Important issues such as sample loading, product cross-contamination, cartridge reuse, and reproducibility are investigated. The SPE system has been demonstrated in the purification of three small libraries that were produced involving amine scavenging reactions with fluorous isatoic anhydride, amide coupling reactions with 2-chloro-4,6-bis [(perfluorohexyl)propyloxy]-1,3,5-triazine (fluorous CDMT), and amide coupling reactions with a newly developed fluorous Mukaiyama condensation reagent.

A 96-well plate-to-plate gravity F-SPE system was designed for solution-phase library purification of four 96-compound libraries produced by scavenging reactions with 1-(perfluoroctyl)propyl isatoic anhydride (F-IA), by amide coupling reactions with 2-chloro-4,6-bis[(perfluoroctyl) propyloxy]-1,3,5-triazine (F-CDMT) or 2,4-dichloro-6-(perfluoroctyl)-propyloxy-1,3,5-triazine (F-DCT), and by Mitsunobu reactions with fluorous diethyl azodicarboxylate (F-DEAD) and triphenylphosphine (F-TPP). Approximately 80% of the products in each library have a purity greater than 85% after F-SPE without conducting chromatographic purification (Zhang et al. 2007).

Using FluoroFlash® cartridges on the RapidTrace workstation, a 10-module workstation was developed to have the capability of completing a maximum of 100 SPEs each round in 1–2 hours (Zhang et al. 2006). Another important feature of the RapidTrace system is that it has the capability of loading slurry samples onto the F-SPE cartridges. The F-SPE cartridge charged with 2 g of fluorous silica gel is used to purify up to 200 mg of crude sample. The automatic SPE system is used for the purification of a small urea library generated from amine-scavenging reactions using fluorous dichlorotriazine, a 96-member amide library generated using 2-chloro-4,6-bis [(perfluorohexyl)propyloxy]-1,3,5-triazine as the coupling agent, and another 96-member library generated from fluorous Mitsunobu reactions. Approximately 90% of the products have a purity >90% after F-SPE.

7.5.3   Fluorous HPLC (F-HPLC)

FluoroFlash® silica gel packed in an HPLC column can be used for separation of mixtures of fluorous compounds based on their different fluorine content (Curran et al. 2001). Molecules with longer fluorine chains (Rf) have longer retention times on the column. A typical mobile phase is a gradient of MeOH–H$_2$O with increasing MeOH up to 100%, which is similar to reversed-phase HPLC. Gradients of MeCN-H$_2$O or THF-H$_2$O can

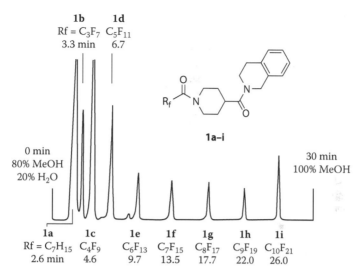

Figure 7.7 Tag-based F-HPLC separation of eight fluorous homologs. (Reproduced with permission from *J. Am. Chem. Soc.* 1999, 121, 9070. Copyright© 1999 American Chemical Society.)

also be used. Figure 7.7 shows a tag-based separation of a 9-component mixture of isonipecotic acid derivatives with different fluorous tags (Rf) (Curran et al. 1999). Eight fluorous compounds (Rf¼C_3F_7 to $C_{10}F_{21}$) were nicely distributed in a 30-minute time period, while the nonfluorous compound (Rf¼C_7H_{15}) was eluted out with the solvent front because of its low affinity with the fluorous stationary phase. F-HPLC analysis and demixing have played a key role in fluorous mixture synthesis.

7.5.4 *Fluorous-flash chromatography (F-FC)*

F-SPE is employed mainly for parallel synthesis. HPLC is good for small-scale purification of the final product, but it has limitation for large-scale intermediate purification. The development of F-FC has provided a scalable separation from 10 mg up to over 10 g by using different sizes of cartridges. Flash chromatography has much better resolution than SPE and is more scalable than HPLC. UV-triggered fraction collection provides reliable fraction cut-off that gives better control on yields and purities than F-SPE. Flash chromatography systems such as those from Biotage (Horizon) and Isco (CombiFlash) are very popular in synthetic labs. These systems equipped with fluorous cartridges can be used for F-FC (Zhang et al. 2004).

7.6 Summary

The combination of compound library synthesis and high-throughput purification has become a powerful tool in today's drug discovery effort. Liquid–liquid extraction, solid-phase capture reagents, and F-based techniques are useful in special cases and for prepurification treatment. However, for large libraries, automated HPLC is a viable alternative. For a robust process, postpurification and informatics support are equally important in the chromatographic separation. Supercritical fluid chromatography has been successfully employed in automated purification instrumentation and is expected to be capable of purifying libraries of tens of thousands of compounds (Ripka et al. 2001).

References

Baldino, C. M. J. 2000. *Comb. Chem.* 2, 89–103.

Berger, T. A., Fogleman, K., Staats, T., Bente, P., Crocket, I., Farrell, W., Osonubi, M. 2000. The development of a semi-preparatory scale supercritical-fluid chromatograph for high-throughput purification of "combi-chem" libraries. *J. Biochem. and Biophys. Meth.* 43(1–3), 87–111.

Blom, K. F., Sparks, R., Doughty, J., J. Everlof, J.G., Haque, T., Combs, A.P. 2003. Optimizing preparative LC/MS configurations and methods for parallel synthesis purification. *J. Comb. Chem.* 5(5), 670–683.

Bookser, B. C., Zhu, S. R. 2001. Solid phase extraction purification of carboxylic acid products from 96-well format solution phase synthesis with DOWEX 1x8-400 formate anion exchange resin. *J. Comb. Chem.* 3(2), 205–215.

Booth, R. J., Hodges, J. C. 1997. Polymer-supported quenching reagents for parallel purification. *J. Am. Chem. Soci.* 119(21), 4882–4886.

Booth, R. J., Hodges, J. C. 1999. Solid-supported reagent strategies for rapid purification of combinatorial synthesis products. *Acc. Chem. Res.* 32(1), 18–26.

Breitenbucher, J. G., Arienti, K.L., McClure, K.J. 2001. Scope and limitations of solid-supported liquid-liquid extraction for the high-throughput purification of compound libraries. *J. Comb. Chem.* 3(6), 528–533.

Cavazzinni, M., Montannari, F., Pozzi, G., Quici, S. J. 1999. *Fluorine Chem.* 94, 183.

Coffey, P. J., Ramieri, J., Garr, C. D., Schultz, L. 1999. *Am. Lab.*

Curran, D. P. 1996. *Chemtracts-Org. Chem.* 9, 75.

Curran, D. P. 1998. *Angew. Chem. Int. Ed. Engl.* 37, 1175.

Curran, D. P. 2001. *Synlett*, 1488.

Curran, D. P., Luo, Z. Y. 1999. Fluorous synthesis with fewer fluorines (light fluorous synthesis): separation of tagged from untagged products by solid-phase extraction with fluorous reverse-phase silica gel. *J. Am. Chem. Soci.* 121(39), 9069–9072.

Curran, D. P., Oderaotoshi, Y. 2001. *Tetrahedron* 57, 5243.

Deal, M. J., Hawes, M. C. 1998. Separation of a two phase mixture by freezing and application to combinatorial libraries. PCT Int. Appl. 20 pp. CODEN: PIXXD2 WO 9851393 A1 19981119 CAN 130:12147 AN 1998:761819.

Edwards, C., Hunter, D.J. 2003. High-throughput purification of combinatorial arrays. *J. Comb. Chem.* 5(1), 61–66.

Espada, A., Molina-Martin, M., Dage, J., Kuo, M. S. 2008. Application of LC/MS and related techniques to high-throughput drug discovery. *Drug Disc. Today* 13(9–10), 417–423.

Farrell, W., Venturea, M., Aurigemma, C., Tran, P., Fiori, K., Xiong, X., Lopez, R., Osbonubi, M. 2001. Analytical and semi-preparative SFC for combinatorial chemistry. Supercritical fluid chromatography, extraction, and processing, Myrtle Beach, SC.

Goetzinger, W., Zhang, X., Bi, G., Towle, M., Cherrak, D., Kyranos, J.N. 2004. High throughput HPLC/MS purification in support of drug discovery. *Int. J. Mass Spectro.* 238, 153–162.

Guth, O., Krewer, D., Freudenberg, B., Paulitz, C., Hauser, M., Ilg, K. 2008. Automated modular preparative HPLC-MS purification laboratory with enhanced efficiency. *J. Comb. Chem.* 10(6), 875–882.

Hochlowski, J., Olson, J., Pan, J., Sauer, D., Searle, P., Sowin, T. 2003. Purification of HTOS libraries by supercritical fluid chromatography. *J. Liq. Chromatogr. Relat. Technol.* 26(3), 333–354.

Hughes, I., Hunter, D. 2001. Techniques for analysis and purification in high-throughput chemistry. *Curr. Opin. Chem. Biol.* (5), 243–247.

Irving, M., Krueger, C.A., Wade, J.V., Hodges, J.C., Leopold, K., Collins, N., Chan, C., Shaqair, S., Shornikov, A., Yan, B. 2004. High-throughput purification of combinatorial libraries II: Automated separation of single diastereomers from a 4-amido-pyrrolidone library containing intentional diastereomer pairs. *J. Comb. Chem.* 6(4), 478–486.

Isbell, J., Xu, R., Cai, Z., Kassel, D.B. 2002. Realities of high-throughput liquid chromatography/mass spectrometry purification of large combinatorial libraries: A report on overall sample throughput using parallel purification. *J. Comb. Chem.* 4(6), 600–611.

Isbell, J. J., Zhou, Y.Y., Guintu, C., Rynd, M., Shumei Jiang, S.M., Petrov, D., Micklash, K., Mainquist, J., Ek, J., Chang, J., Weselak, M., Backes, B.J., Brailsford, A., Shave, D. 2005. Purifying the masses: integrating pre-purification quality control, high-throughput LC/MS purification, and compound plating to feed high-throughput screening. *J. Comb. Chem.* 7(2), 210–217.

Johnson, C. R., Zhang, B. R., Fantauzzi, P., Hocker, M., Yager, K. M. 1998. Libraries of N-alkylaminoheterocycles from nucleophilic aromatic substitution with purification by solid supported liquid extraction. *Tetrahedron* 54(16), 4097–4106.

Kaldor, S. W., Siegel, M. G., Fritz, J. E., Dressman, B. A., Hahn, P. J. 1996. Use of solid supported nucleophiles and electrophiles for the purification of non-peptide small molecule libraries. *Tetra. Lett.* 37(40), 7193–7196.

Karancsi, T., Goedoerhazy, L., Szalay, D., Darvas, F. 2005a. UV-triggered main-component fraction collection method and its application for high-throughput chromatographic purification of combinatorial libraries. *J. Comb. Chem.* 7(1), 58–62.

Karancsi, T., Goedoerhzy, L., Szalay, D., Papp, B., Nemeth, L., Darvas, F. 2005b. Application of main component fraction collection method for purification of compound libraries. *J. Chromatogr. A* 1079, 349–353.

Koppitz, M., Brailsford, A., Wenz, M. 2005. Maximizing automation in LC/MS high-throughput analysis and purification. *J. Comb. Chem.* 7(5), 714–720.

Leister, W., Strauss, K., Wisnoski, D., Zhao, Z.J., Lindsley, C. 2003. Development of a custom high-throughput preparative liquid chromatography/mass spectrometer platform for the preparative purification and analytical analysis of compound libraries. *J. Comb. Chem.* 5(3), 322–329.

Li, F. B., Hsieh, Y. S. 2008. Supercritical fluid chromatography-mass spectrometry for chemical analysis. *J. Separ. Sci.* 31(8), 1231–1237.

Mukherjee, S., Poon, K. W. C., Flynn, D. L., Hanson, P. R. 2003. Capture-ROMP-release: Application to the synthesis of amines and alkyl hydrazines. *Tetra. Lett.* 44(38), 7187–7190.

Nilsson, U. J. 2000. Solid-phase extraction for combinatorial libraries. *J. Chromatogr. A* 885(1–2), 305–319.

Olson, J., Pan, J., Hochlowski, J., Searle, P., Blanchard, D. 2002. Customization of a commercially available prep scale SFC system to provide enhanced capabilities. *JALA* 7(4), 69–74.

Parlow, J. J., Devraj, R. V., South, M. S. 1999. Solution-phase chemical library synthesis using polymer-assisted purification techniques. *Curr. Opin. Chem. Biol.* 3(3), 320–336.

Peng, S. X., Henson, C., Strojnowski, M.J., Golebiowski, A., Klopfenstein, S.R. 2000. Automated high-throughput liquid-liquid extraction for initial purification of combinatorial libraries. *Anal. Chem.* 72(2), 261–266.

Pirrung, M. C., Tumey, L. N., McClerren, A. L., Raetz, C. R. H. 2003. High-throughput catch-and-release synthesis of oxazoline hydroxamates. Structure-activity relationships in novel inhibitors of Escherichia coli LpxC: In vitro enzyme inhibition and antibacterial properties. *J. Am. Chem. Soc.* 125(6), 1575–1586.

Ripka, W. C., Barker, G., Krakover, J. 2001. High-throughput purification of compound libraries. *Drug Disc. Today* 6(9), 471–477.

Schaffrath, M., von Roedern, E., Hamley, P., Stilz, H.U. 2005. High-throughput purification of single compounds and libraries. *J. Comb. Chem.* 7(4), 546–553.

Schultz, L., Garr, C. D., Cameron, L. M., Bukowski, J. 1998. *Bioorg. Med. Chem. Lett.* 8, 2409–2414.

Searle, P. A., Glass, K.A., Hochlowski, J.E. 2004. Comparison of preparative HPLC/MS and preparative SFC techniques for the high-throughput purification of compound libraries. *J. Comb. Chem.* 6(2), 175–180.

Studer, A., Hadida, S., Ferritto, S. Y., Kim, P. Y., Jeger, P., Wipf, P., Curran, D. P. 1997. Fluorous synthesis: A fluorous-phase strategy for improving separation efficiency in organic synthesis. *Science* 275(5301), 823–826.

Taylor, L. T. 2008. Supercritical fluid chromatography. *Anal. Chem.* 80(12), 4285–4294.

Thompson, L. A. 2000. Recent applications of polymer-supported reagents and scavengers in combinatorial, parallel, or multistep synthesis. *Curr. Opin. Chem. Biol.* 4(3), 324–337.

Tzschucke, C. C., Markert, C., Bannwarth, W., Roller, S., Hebel, A., Haag, R. Angew. 2002. *Chem. Int. Ed. Engl.* 41, 3964.

Wang, T., Barber, M., Hardt, I., Kassel, D. B. 2001. Mass-directed fractionation and isolation of pharmaceutical compounds by packed-column supercritical fluid chromatography/mass spectrometry. *Rapid Commun. Mass Spectrom.* 15, 2067–2075.

Weller, H. N. 1999. Purification of combinatorial libraries. *Mol. Diversity* 4, 47–52.

Weller, H. N., Young, M. G., Michalczyk, S. J., Reitnauer, G. H., Cooley, R. S., Rahn, P. C., Loyd, D. J., Fiore, D., Fischman, S. J. 1997. High throughput analysis and purification in support of automated parallel synthesis. *Mol. Diversity* 3(1), 61–70.

Xu, R., Wang, T., Isbell, J., Cai, Z., Sykes, C., Brailsford, A., Kassel, D.B. 2002. High-throughput mass-directed parallel purification incorporating a multiplexed single quadrupole mass spectrometer. *Anal. Chem.* 74(13), 3055–3062.

Yan, B. 2003. *Analysis and Purification Methods in Combinatorial Chemistry.* John Wiley and Sons, Inc,.

Yan, B., Collins, N., Wheatley, J., Irving, M., Leopold, K., Chan, C., Shornikov, A., Fang, L.L., Lee, A., Stock, M., Zhao, J. 2004. High-throughput purification of combinatorial libraries I: A high-throughput purification system using an accelerated retention window approach. *J. Comb. Chem.* 6(2), 255–261.

Yan, B., Czarnik, A. W. 2001. *Optimization of Solid-Phase Combinatorial Synthesis.* New York, Marcel Dekker.

Yan, B., Fang, L., Irving, M., Zhang, S., Boldi, A., Woolard, F., Figliozzi, G. M., Lease, T. G., Johnson, C. R., Collins, N. 2003. Quality control in combinatorial chemistry: Determination of the quantity, purity, and quantitative purity of compounds in combinatorial libraries. *J. Comb. Chem.* 5(5), 547–559.

Zeng, L., Kassel, D.B. 1998. Developments of a fully automated parallel HPLC/ Mass spectrometry system for the analytical characterization and preparative purification of combinatorial libraries. *Anal. Chem.* 70(20), 4380–4388.

Zhang, W. 2003. Fluorous technologies for solution-phase high-throughput organic synthesis. *Tetrahedron* 59, 4475–4489.

Zhang, W. 2004. Fluorous synthesis of heterocyclic systems. *Chem. Rev.* 104(5), 2531–2556.

Zhang, W., Curran, D.P. 2006. Synthetic applications of fluorous solid-phase extraction (F-SPE). *Tetrahedron* 62, 11837–11865.

Zhang, W., Lu, Y. 2006. Automation of fluorous solid-phase extraction for parallel synthesis. *J. Comb. Chem.* 8(6), 890–896.

Zhang, W., Lu, Y.M. 2007. 96-well plate-to-plate gravity fluorous solid-phase extraction (F-SPE) for solution-phase library purification. *J. Comb. Chem.* 9(5), 836–843.

Zhang, W., Lu, Y.M., Nagashima, T. 2005. Plate-to-plate fluorous solid-phase extraction for solution-phase parallel synthesis. *J. Comb. Chem.* 7(6), 893–897.

Zhang, X., Picariello, W., Hosein, N., Towle, M., Goetzinger, W. 2006. A systematic investigation of recovery in preparative reverse phase high performance liquid chromatography/mass spectrometry. *J. Chromatogr. A* 1119(1–2), 147–155.

Zhang, X., Towle, M.H., Felice, C.E., Flament, J.H., Goetzinger, W.K. 2006. Development of a mass-directed preparative supercritical fluid chromatography purification system. *J. Comb. Chem.* 8(5), 705–714.

chapter eight

Final thoughts and future perspectives

8.1 Introduction

Previous chapters have provided an overview of the major analytical technologies in combinatorial chemistry. Examples and highlights of many recently published results have shown that tremendous progress has already been made in this area. These citations are by no means complete, but they do represent where we are. Since summaries on each methodology have been given at the end of each chapter, this chapter will suggest only some remaining issues that need to be solved and trends for future developments.

In the first edition, we made the prediction that miniaturized and parallel analysis techniques were needed and would be developed. Such technologies have in fact been extensively developed in the past decade. We witnessed the development of nano LC, MUX technology for LC/MS, UPLC/MS, flow chemistry, 96-channel CE, and various parallel miniaturized ADME profiling assays. We will further point out some representative areas where breakthroughs are expected.

8.2 Stability of compound collections

Drug lead discovery heavily depends on the screening of diverse organic compounds against validated therapeutic targets using high-throughput screening (HTS) technologies. Compound diversity is an important element in building such compound collections. However, diverse compound structures and crystallinity and the associated physicochemical properties have caused tremendous problems in compound solubility, stability, and the overall quality of the collection. The quality of compound collections has an immense impact on the credibility of screening results and the advancement of drug candidates, and, therefore, on the drug discovery timeline. Because compound decomposition is unavoidable, one should be able to determine compound purity and quantity. The standard method for quantitation of organic compounds is evidently not suitable for the inhibitory large number of compounds in any compound collection (Qi et al. 2008). Quantification methods of trace amounts of organic compounds with high-throughput will be developed. The data mining and correlation to regular structure will be important and will be established using cheminformatics techniques.

8.3 Nanocombinatorial library and associated analytical issues

Nanotechnology is a rapidly progressing field focused on the creation of functional materials, devices, and systems through the control of matter on the nanometer scale. The successful application of combinatorial chemistry in the discovery of new chemical entities, such as novel drugs and catalysts, will also have a tremendous impact on techniques used in nanotechnology research. Functional and multifunctional nanoparticles are becoming powerful materials in nanomedicine, and diverse chemical research on nanoparticle surfaces using combinatorial chemistry has brought about nanomaterial biocompatability and specificity (Mu et al. 2008; Zhang et al. 2009). A new concept of *nano-combinatorial library* that aims to accelerate the discovery of novel nanomaterials with unique properties has been developed (Zhou et al. 2008). More nanocombinatorial chemical research is unavoidable.

An important issue is developing an analytical strategy for characterizing surface-modified nanoparticles from the nanocombinatorial library (Zhang and Yan 2009). This is very challenging because these solid-phase samples are usually coated with only a small number of molecules. More efforts will be made to fill this gap. For example, 1D and 2D high-resolution magnetic angle spinning NMR (Zhou et al. 2008, 2009; Du et al. 2009), FTIR, and MS techniques will be also used to monitor the reaction conversions in the multistep modifications (Fleming et al. 2006; Tan et al. 2006; Male et al. 2008). X-ray photoelectron spectroscopy (XPS) and combustion elemental analysis are techniques for determining the chemical compositions of molecules bound to NPs (Holzinger et al. 2003; Kumar et al. 2003; Bruce and Sen 2005; del Campo et al. 2005; Zhou et al. 2008). LC-MS analysis after ligand cleavage is the only method for qualitative and quantitative analysis of multiple functionalized NPs when a quantitative detector such as a CLND is used.

Although significant progress has been made in analytical strategies for nanomaterial surface chemistry, identifying the attached molecules and quantifying their loading still require our continued dedication and efforts.

8.4 Green process and technologies

In recent years, the advancement of green chemistry has been showcased by numerous examples of large-scale process chemistry and chemical engineering. This will become equally important in combinatorial library synthesis. Using environmentally friendly solvents and renewable feed-stocks and developing chemical recycling techniques have become common practices in the chemical industry (Lankey and Anastas 2005). At the same time, many new tools with the potential for green chemistry applications have been introduced during the development of combinatorial chemistry. Among them, fluorous chemistry overlaps with many combinatorial techniques (Zhang 2009), and other green techniques in reaction, analysis, and separation will be developed soon.

References

Bruce, I. J., and Sen, T. 2005. Surface modification of magnetic nanoparticles with alkoxysilanes and their application in magnetic bioseparations. *Langmuir* 21, 7029–7035.

del Campo, A., Sen, T., Lellouche, J.-P., Bruce, I. J. 2005. Multifunctional magnetite and silica–magnetite nanoparticles: Synthesis, surface activation and applications in life sciences. *J. Magn. Mater.* 293, 33–40.

Du, F.F., Zhou, H.Y., Chen, L.X., Zhang, B., Yan, B. 2009. Structure elucidation of nanoparticle-bound organic molecules by ^{1}H NMR. *Trends Anal. chem.* 28, 88–95.

Fleming, D.A., Thode, C. J., Williams, M. E. 2006. Triazole cycloaddition as a general route for functionalization of au nanoparticles. *Chem. Mater.* 18, 2327–2334.

Holzinger, M., Abraham, J., Whelan, P., Graupner, R., Ley, L., Hennrich, F., Kappes, M., Hirsch, A. 2003. Functionalization of single-walled carbon nanotubes with (R-)oxycarbonyl nitrenes. *J. Am. Chem. Soc.* 125, 8566–8580.

Kumar, A., Mandal, S., Selvakannan, P. R., Pasricha, R., Mandale, A. B., Sastry, M. 2003. Investigation into the interaction between surface-bound alkylamines and gold nanoparticles. *Langmuir* 19, 6277–6282.

Lankey, R.L. and Anastas, P. T. (eds.) 2005. *Advancing Sustainability Through Green Chemistry and Engineering.* American Chemical Society Publication.

Male, K.B., Li, J., Bun, C.C., Ng, S.-C., Luong, J.H.T. 2008. Synthesis and stability of fluorescent gold nanoparticles by sodium borohydride in the presence of mono-6-deoxy-6-pyridinium-â-cyclodextrin chloride. *J. Phys. Chem.* C 112, 443–451.

Mu, Q.X., Liu, W., Xing, Y.H., Zhou, H.Y., Li, Z.W., Zhang, Y., Ji, L.H., Wang, F., Si, Z.K., Zhang, B., Yan, B. 2008. Protein binding by functionalized multi-walled carbon nanotubes is governed by the surface chemistry of both parties and the nanotube diameter. *J. Phys. Chem.* C 112, 3300–3307.

Qi, M.H., Zhou, H.Y., Ma, X.F., Zhang, B., Jefferies, C., Yan, B. 2008. Feasibility of a self-calibrated LC/MS/UV method to determine the absolute amount of compounds in their storage and screening lifecycle. *J. Comb. Chem.* 10, 162–165.

Tan, H., Zhan, T., Fan, W.Y. 2006. Direct functionalization of the hydroxyl group of the 6-Mercapto-1-hexanol (MCH) ligand attached to gold nanoclusters. *J. Phys. Chem.* B 110, 21690–21693.

Zhang, B., Li, Z.W., Xing, Y.H., Zhou, H.Y., Mu, Q.X., Yan, B. 2009. Functionalized carbon nanotubes specifically bind to chymotrypsin catalytic site and regulate its enzymatic function. *Nano. Lett.* 9, 2280–2284.

Zhang, B. and Yan, B. 2010. Analytical strategies for characterizing the surface chemistry of nanoparticles. *Anal. Bioanal. Chem.* 396, 973–982.

Zhang, W. 2009. Green chemistry aspects of fluorous techniques-opportunities and challenges for small-scale organic synthesis. *Green Chem.* 11, 911–920.

Zhou, H. Y., Du, F. F., Li, X., Zhang, B., Li, W., Yan, B. 2008. Characterization of organic molecules attached to gold nanoparticle surface using high resolution magic angle spinning ¹H NMR. *J. Phys. Chem.* C 112, 19360–19366.

Zhou, H.Y., Mu, Q.X., Gao, N.N., Liu, A.F., Xing, Y.H., Gao, S.L., Zhang, Q., Qu, G.B., Chen, Y.Y., Liu, G., Zhang, B., Yan, B. 2008. A nano-combinatorial library strategy for the discovery of nanotubes with reduced protein-binding, cytotoxicity, and immune response. *Nano. Letters* 8, 859–865.

Index